DATE DUE			
Dec 11 '73			
Apr 12 '76			
Aug 9 '77			
Apr 12 '78			
Aug 14 '83			
Feb 23 '81			
Dec 10 '82			

THE ABC OF RELATIVITY

BY BERTRAND RUSSELL

The Analysis of Matter
Human Society in Ethics and Politics
The Impact of Science on Society
New Hopes for a Changing World
Authority and the Individual
Human Knowledge: its Scope and Limits
History of Western Philosophy
The Principles of Mathematics
Introduction to Mathematical Philosophy
The Analysis of Mind
Our Knowledge of the External World
An Outline of Philosophy
The Philosophy of Leibniz
An Inquiry into Meaning and Truth
Logic and Knowledge
The Problems of Philosophy
Principia Mathematica

Fact and Fiction
Has Man a Future
Unarmed Victory
My Philosophical Development
Why I am not a Christian
Common Sense and Nuclear Warfare
Portraits from Memory
Unpopular Essays
Power
In Praise of Idleness
The Conquest of Happiness
Sceptical Essays
Mysticism and Logic
The Scientific Outlook
Marriage and Morals
Education and the Social Order
On Education

Freedom and Organisation, 1814-1914
Principles of Social Reconstruction
Roads to Freedom
Practice and Theory of Bolshevism

Satan in the Suburbs
Nightmares of Eminent Persons

★

Bertrand Russell's Best
Selected and Introduced
by ROBERT E. EGNER

★

The Prospects of Industrial Civilization
by BERTRAND RUSSELL and DORA RUSSELL

BERTRAND RUSSELL

The ABC of Relativity

Third Revised Edition

EDITED BY FELIX PIRANI

Ruskin House
GEORGE ALLEN & UNWIN LTD
MUSEUM STREET LONDON

First Published 1925
Reprinted 1925, 1927, 1931
Second Edition 1958
Second Impression 1959
Third Impression 1964
Third Edition 1969

SBN 04 521001 2

PRINTED IN GREAT BRITAIN
by Photolithography
JOHN DICKENS AND CO LTD, NORTHAMPTON

NOTE

This book first appeared in 1925. The basic principles of relativity have not changed since then, but its applications have been much extended, and some revision was necessary for the second edition, which was carried out, with my approval, by Mr. Felix Pirani. For this third edition Mr. Pirani has further revised a number of passages to agree with present knowledge.

B.R.

CONTENTS

CHAPTER I

Touch and Sight: The Earth and the Heavens

EVERYBODY knows that Einstein did something astonishing, but very few people know exactly what it was that he did. It is generally recognized that he revolutionized our conception of the physical world, but the new conceptions are wrapped up in mathematical technicalities. It is true that there are innumerable popular accounts of the theory of relativity, but they generally cease to be intelligible just at the point where they begin to say something important. The authors are hardly to blame for this. Many of the new ideas can be expressed in non-mathematical language, but they are none the less difficult on that account. What is demanded is a change in our imaginative picture of the world—a picture which has been handed down from remote, perhaps pre-human, ancestors, and has been learned by each one of us in early childhood. A change in our imagination is always difficult, especially when we are no longer young. The same sort of change was demanded by Copernicus, when he taught that the earth is not stationary and the heavens do not revolve about it once a day. To us now there is no difficulty in this idea, because we learned it before our mental habits had become fixed. Einstein's ideas, similarly, will seem easier to generations which grow up with them; but for us a certain effort of imaginative reconstruction is unavoidable.

In exploring the surface of the earth, we make use of all our senses, more particularly of the senses of touch and sight. In measuring lengths, parts of the human body are employed in pre-scientific ages: a 'foot,' a 'cubit,' a 'span' are defined in this way. For longer distances, we think of the time it takes to walk from one place to another. We gradually learn to judge distance roughly by the eye, but we rely upon touch for accuracy.

Moreover it is touch that gives us our sense of 'reality.' Some things cannot be touched: rainbows, reflections in looking-glasses, and so on. These things puzzle children, whose metaphysical speculations are arrested by the information that what is in the looking-glass is not 'real.' Macbeth's dagger was unreal because it was not 'sensible to feeling as to sight.' Not only our geometry and physics, but our whole conception of what exists outside us, is based upon the sense of touch. We carry this even into our metaphors: a good speech is 'solid,' a bad speech is 'gas,' because we feel that a gas is not quite 'real.'

In studying the heavens, we are debarred from all senses except sight. We cannot touch the sun, or travel to it; we cannot yet walk round the moon, or apply a foot-rule to the Pleiades. Nevertheless, astronomers have unhesitatingly applied the geometry and physics which they found serviceable on the surface of the earth, and which they had based upon touch and travel. In doing so, they brought down trouble on their heads, which it was left for Einstein to clear up. It turned out that much of what we learned from the sense of touch was unscientific prejudice, which must be rejected if we are to have a true picture of the world.

An illustration may help us to understand how much is impossible to the astronomer as compared with the man who is interested in things on the surface of the earth. Let us suppose that a drug is administered to you which makes you temporarily unconscious, and that when you wake you have lost your memory but not your reasoning powers. Let us suppose further that while you were unconscious you were carried into a balloon, which, when you come to, is sailing with the wind on a dark night—the night of the fifth of November if you are in England, or of the fourth of July if you are in America. You can see fireworks which are being sent off from the ground, from trains, and from aeroplanes travelling in all directions, but you cannot see the ground or the trains or the aeroplanes because of the darkness. What sort of picture of the world will you form? You will think that nothing is permanent: there are only brief flashes of light, which, during their short existence, travel through the void in the most various and bizarre curves. You cannot touch these flashes of light, you can only see them. Obviously your geometry and your physics and your metaphysics will be quite different from those of ordinary mortals. If an ordinary mortal were with you in the balloon, you would find his speech unintelligible. But if Einstein were with you, you would understand him more easily than the ordinary mortal would, because you would be free from a host of preconceptions which prevent most people from understanding him.

The theory of relativity depends, to a considerable extent, upon getting rid of notions which are useful in ordinary life but not to our drugged balloonist. Circumstances on the surface of the earth, for various more or less accidental reasons, suggest conceptions which turn out to be inaccurate, although they have come to seem like necessities of thought. The most important of these circumstances is that most objects on the earth's surface are fairly persistent and nearly stationary from a terrestrial point of view. If this were not the case, the idea of going on a journey would not seem so definite as it does. If you want to travel from King's Cross to Edinburgh, you know that you will find King's Cross where it has always been, that the railway line will take the course that it did when you last made the journey, and that Waverley Station in Edinburgh will not have walked up to the Castle. You therefore say and think that you have travelled to Edinburgh, not that Edinburgh has travelled to you, though the latter statement would be just as accurate. The success of this common-sense point of view depends upon a number of things which are really of the nature of luck. Suppose all the houses in London were perpetually moving about, like a swarm of bees; suppose railways moved and changed their shapes like avalanches; and finally suppose that material objects were perpetually being formed and dissolved like clouds. There is nothing impossible in these suppositions. But obviously what we call a journey to Edinburgh would have no meaning in such a world. You would begin, no doubt, by asking the taxi-driver: 'Where is King's Cross this morning?' At the station you would have to ask a similar question about Edinburgh, but the booking-office clerk would reply: 'What part of Edinburgh do you mean, sir? Prince's Street has gone to Glasgow, the Castle has moved up into the Highlands, and Waverley Station is under water in the middle of the Firth of Forth.' And on the journey the stations would not be staying quiet, but some would be travelling north, some south, some east or west, perhaps much faster than the train. Under these conditions you could not say where you were at any moment. Indeed the whole notion that one is always in some definite 'place' is due to the fortunate immobility of most of the large objects on the earth's surface. The idea of 'place' is only a rough practical approximation: there is nothing logically necessary about it, and it cannot be made precise.

If we were not much larger than an electron, we should not have this impression of stability, which is only due to the grossness of our senses. King's Cross, which to us looks solid, would be too vast to be conceived except by a few eccentric mathematicians. The bits of it that we could see would consist of little tiny points of matter, never coming into contact with each other, but perpetually whizzing round each other in an inconceivably rapid ballet-dance. The world of our experience would be quite as mad as the one in which the different parts of Edinburgh go for walks in different directions. If—to take the opposite extreme—you were as large as the sun and lived as long, with a corresponding slowness of perception, you would again find a higgledy-piggledy universe without permanence—stars and planets would come and go like morning mists, and nothing would remain in a fixed position relatively to anything else. The notion of comparative stability which forms part of our ordinary outlook is thus due to the fact that we are about the size we are, and live on a planet of which the surface is not very hot. If this were not the case, we should not find pre-relativity physics intellectually satisfying. Indeed we should never have invented such theories. We should have had to arrive at relativity physics at one bound, or remain ignorant of scientific laws. It is fortunate for us that we were not faced with this alternative, since it is almost inconceivable that one man could have done the work of Euclid, Galileo, Newton and Einstein. Yet without such an incredible genius physics could hardly have been discovered in a world where the universal flux was obvious to non-scientific observation.

In astronomy, although the sun, moon, and stars continue to exist year after year, yet in other respects the world we have to deal with is very different from that of everyday life. As already observed, we depend exclusively on sight: the heavenly bodies cannot be touched, heard, smelt or tasted. Everything in the heavens is moving relatively to everything else. The earth is going round the sun, the sun is moving, very much faster than an express train, towards a point in the constellation Hercules, the 'fixed' stars are scurrying hither and thither like a lot of frightened hens. There are no well-marked places in the sky, like King's Cross and Edinburgh. When you travel from place to place on the earth, you say the train moves and not the stations, because the stations preserve their topographical relations to each other and the surrounding country. But in astromony it is arbitrary which you call the train and which the station: the question is to be decided purely by convenience and as a matter of convention.

In this respect, it is interesting to contrast Einstein and Copernicus. Before Copernicus, people thought that the earth stood still and the heavens revolved about it once a day. Copernicus taught that 'really' the earth rotates once a day, and the daily revolution of sun and stars is only 'apparent.' Galileo and Newton endorsed this view, and many things were thought to prove it—for example, the flattening of the earth at the poles, and the fact that bodies are heavier there than at the equator. But in the modern theory the question between Copernicus and his predecessors is merely one of convenience; all motion is relative, and there is no difference between the two statements: 'the earth rotates once a day' and 'the heavens revolve about the earth once a day.' The two mean exactly the same thing, just as it means the same thing if I say that a certain length is six feet or two yards. Astronomy is easier if we take the sun as fixed than if we take the earth, just as accounts are easier in decimal coinage. But to say more for Copernicus is to assume absolute motion, which is a fiction. All motion is relative, and it is a mere convention to take one body as at rest. All such conventions are equally legitimate, though not all are equally convenient.

There is another matter of great importance, in which astronomy differs from terrestrial physics because of its exclusive dependence upon sight. Both popular thought and old-fashioned physics used the notion of 'force,' which seemed intelligible because it was associated with familiar sensations. When we are walking, we have sensations connected with our muscles which we do not have when we are sitting still. In the days before mechanical traction, although people could travel by sitting in their carriages, they could see the horses exerting themselves, and evidently putting out 'force' in the same way as human beings do. Everybody knew from experience what it is to push or pull, or to be pushed or pulled. These very familiar facts made 'force' seem a natural basis for dynamics. But Newton's law of gravitation introduced a difficulty. The force between two billiard balls appeared intelligible because we know what it feels like to bump into another person; but the force between the earth and the sun, which are ninety-three million miles apart, was mysterious. Newton himself regarded this 'action at a distance' as impossible, and believed that there was some hitherto undiscovered mechanism by which the sun's influence was transmitted to the planets. However, no such mechanism was discovered, and gravitation remained a puzzle. The fact is that the whole conception of 'force' is a mistake. The sun does not exert any force on the planets; in Einstein's law of gravitation, the planet only pays attention to what it finds in its own neighbourhood. The way in which this works will be explained in a later chapter; for the present we are only concerned with the necessity of abandoning the notion of 'force,' which was due to misleading conceptions derived from the sense of touch.

As physics has advanced, it has appeared more and more that sight is less misleading than touch as a source of fundamental notions about matter. The apparent simplicity in the collision of billiard balls is quite illusory. As a matter of fact the two billiard balls never touch at all; what really happens is inconceivably complicated, but is more analogous to what happens when a comet penetrates the solar system and goes away again than to what common sense supposes to happen.

Most of what we have said hitherto was already recognized by physicists before Einstein invented the theory of relativity. 'Force' was known to be merely a mathematical fiction, and it was generally held that motion is a merely relative phenomenon —that is to say, when two bodies are changing their relative position, we cannot say that one is moving while the other is at rest, since the occurrence is merely a change in their relation to each other. But a great labour was required in order to bring the actual procedure of physics into harmony with these new convictions. Newton believed in force and in absolute space and time; he embodied these beliefs in his technical methods, and his methods remained those of later physicists. Einstein invented a new technique, free from Newton's assumptions. But in order to do so he had to change fundamentally the old ideas of space and time, which had been unchallenged from time immemorial. This is what makes both the difficulty and the interest of his theory. But before explaining it there are some preliminaries which are indispensable. These will occupy the next two chapters.

CHAPTER II

What Happens and What is Observed

A CERTAIN type of superior person is fond of asserting that 'everything is relative.' This is, of course, nonsense, because, if *everything* were relative, there would be nothing for it to be relative to. However, without falling into metaphysical absurdities it is possible to maintain that everything in the physical world is relative to an observer. This view, true or not, is *not* that adopted by the 'theory of relativity.' Perhaps the name is unfortunate; certainly it has led philosophers and uneducated people into confusions. They imagine that the new theory proves *everything* in the physical world to be relative, whereas, on the contrary, it is wholly concerned to exclude what is relative and arrive at a statement of physical laws that shall in no way depend upon the circumstances of the observer. It is true that these circumstances have been found to have more effect upon what appears to the observer than they were formerly thought to have, but at the same time Einstein showed how to discount this effect completely. This was the source of almost everything that is surprising in his theory.

When two observers perceive what is regarded as one occurrence, there are certain similarities, and also certain differences, between their perceptions. The differences are obscured by the requirements of daily life, because from a business point of view they are as a rule unimportant. But both psychology and physics, from their different angles, are compelled to emphasize the respects in which one man's perception of a given occurrence differs from another man's. Some of these differences are due to differences in the brains or minds of the observers, some to differences in their sense-organs, some to differences of physical situation: these three kinds may be called respectively psychological, physiological, and physical. A remark made in a language we know will be heard, whereas an equally loud remark in an unknown language may pass entirely unnoticed. Of two men in the Alps, one will perceive the beauty of the scenery while the other will notice the waterfalls with a view to obtaining power from them. Such differences are psychological. The differences between a long-sighted and a short-sighted man, or between a deaf man and a man who hears well, are physiological. Neither of these kinds concerns us, and I have mentioned them only in order to exclude them. The kind that concerns us is the purely physical kind. Physical differences between two observers will be preserved when the observers are replaced by cameras or recording machines, and can be reproduced in a film or on the gramophone. If two men both listen to a third man speaking, and one of them is nearer to the speaker than the other is, the nearer one will hear louder and slightly earlier sounds than are heard by the other. If two men both watch a tree falling, they see it from different angles. Both these differences would be shown equally by recording instruments: they are in no way due to idiosyncrasies in the observers, but are part of the ordinary course of physical nature as we experience it.

The physicist, like the plain man, believes that his perceptions give him knowledge about what is really occurring in the physical world, and not only about his private experiences. Professionally, he regards the physical world as 'real,' not merely as something which human beings dream. An eclipse of the sun, for instance, can be observed by any person who is suitably situated, and is also observed by the photographic plates that are exposed for the purpose. The physicist is persuaded that something has really happened over and above the experience of those who have looked at the sun or at photographs of it. I have emphasized this point, which might seem a trifle obvious, because some people imagine that Einstein made a difference in this respect. In fact he has made none.

But if the physicist is justified in this belief that a number of people can observe the 'same' physical occurrence, then clearly the physicist must be concerned with those features which the occurrence has in common for all observers, for the others cannot be regarded as belonging to the occurrence itself. At least, the physicist must confine himself to the features which are common to all 'equally good' observers. The observer who uses a microscope or a telescope is preferred to one who does not, because he sees all that the latter sees and more too. A sensitive photographic plate may 'see' still more, and is then preferred to any eye. But such things as differences of perspective, or differences of apparent size, due to difference of distance, are obviously not attributable to the object; they belong solely to the point of view of the spectator. Common sense eliminates these in judging of objects; physics has to carry the same process much further, but the principle is the same.

I want to make it clear that I am not concerned with anything that can be called inaccuracy. I am concerned with genuine physical differences between occurrences each of which is a correct record of a certain event, from its own point of view. When a man fires a gun, people who are not quite close to him see the flash before they hear the report. This is not due to any defect in their senses, but to the fact that sound travels more slowly than light. Light travels so fast that, from the point of phenomena on the surface of the earth, it may be regarded as instantaneous. Anything that we can see on the earth happens practically at the moment when we see it. In a second, light travels 300,000 kilometres (about 186,000 miles). It travels from the sun to the earth in about eight minutes, and from the stars in anything from four years to several thousand million. But of course we cannot place a clock on the sun, and send out a flash of light from it at 12 noon, Greenwich Mean Time, and have it received at Greenwich at 12.8 p.m. Our methods of estimating the speed of light have to be more or less indirect. The most direct method is that which we apply to sound when we use an echo. We could send a flash to a mirror, and observe how long it took for the reflection to reach us; this would give the time of the double journey to the mirror and back. On the earth, however, this time would be inconveniently short, so that in practice the physicists have to use a more complicated method, but the underlying principle is still that of the echo.

The same principle is used, for another purpose, in radar. Very short radio waves (whose speed is the same as the speed of light) are sent out and reflected back from a distant object. Then the distance of the object can be deduced from the time it takes the waves to go there and back.

The problem of allowing for the spectator's point of view, we may be told, is one of which physics has at all times been fully aware; indeed it has dominated astronomy ever since the time of Copernicus. This is true. But principles are often acknowledged long before their full consequences are drawn. Much of traditional physics is incompatible with the principle, in spite of the fact that it was acknowledged theoretically by all physicists.

There existed a set of rules which caused uneasiness to the philosophically minded, but were accepted by physicists because they worked in practice. Locke had distinguished 'secondary' qualities—colours, noises, tastes, smells, etc.—as subjective, while allowing 'primary' qualities—shapes and positions and sizes—to be genuine properties of physical objects. The physicist's rules were such as would follow from this doctrine. Colours and noises were allowed to be subjective, but due to waves proceeding with a definite velocity—that of light or sound as the case may be—from their source to the eye or ear of the percipient. Apparent shapes vary according to the laws of perspective, but these laws are simple and make it easy to infer the 'real' shapes from several visual apparent shapes; moreover, the 'real' shapes can be ascertained by touch in the case of bodies in our neighbourhood. The objective time of a physical occurrence can be inferred from the time when we perceive it by allowing for the velocity of transmission —of light or sound or nerve currents according to circumstances. This was the view adopted by physicists in practice, whatever qualms they may have had in unprofessional moments.

This view worked well enough until physicists became concerned with much greater velocities than those that are common on the surface of the earth. An express train travels about a mile in a minute; the planets travel a few miles in a second. Comets, when they are near the sun, travel much faster, but because of their continually changing shapes it is impossible to determine their positions very accurately. Practically, the planets were the most swiftly-moving bodies to which dynamics could be adequately applied. With the discovery of radioactivity and cosmic rays, and recently with the construction of high energy accelerating machines, new ranges of observation have become possible. Individual sub-atomic particles can be observed, moving with velocities not far short of that of light. The behaviour of bodies moving with these enormous speeds is not what the old theories would lead us to expect. For one thing, mass seems to increase with speed in a perfectly definite manner. When an electron is moving very fast, a bigger force is required to have a given effect upon it than when it is moving slowly. Then reasons have been found for thinking that the size of a body is affected by its motion—for example, if you take a cube and move it very fast, it gets shorter in the direction of its motion, from the point of view of a person who is not moving with it, though from its its own point of view (i.e. for an observer travelling with it) it remains just as it was. What was still more astonishing was the discovery that lapse of time depends on motion; that is to say, two perfectly accurate clocks, one of which is moving very fast relatively to the other, will not continue to show the same time if they come together again after a journey. This is too small an effect to have been tested directly so far, but it should be possible to test it if we ever succeed in developing space travel, for then we shall be able to make journeys long enough for this 'time dilatation,' as it is called, to become quite appreciable.

There is some direct evidence for the time dilatation, but it is found in a different way. This evidence comes from observations of cosmic rays, which consist of a variety of atomic particles coming from outer space and moving very fast through the earth's atmosphere. Some of these particles, called mesons, disintegrate in flight, and the disintegration can be observed. It is found that the faster a meson is moving, the longer it takes to disintegrate, from the point of view of a scientist on the earth. It follows from results of this kind that what we discover by means of clocks and foot-rules, which used to be regarded as the acme of impersonal science, is really in part dependent upon our private circumstances, i.e. upon the way in which we are moving relatively to the bodies measured.

This shows that we have to draw a different line from that which is customary in distinguishing between what belongs to the observer and what belongs to the occurrence which he is observing. If a man is wearing blue spectacles he knows that the blue look of everything is due to his spectacles, and does not belong to what he is observing. But if he observes two flashes of lightning, and notes the interval of time between his observations; if he knows where the flashes took place, and allows, in each case, for the time the light took to reach him—in that case, if his chronometer is accurate, he naturally thinks that he has discovered the actual interval of time between the two flashes, and not something merely personal to himself. He is confirmed in this view by the fact that all other careful observers to whom he has access agree with his estimates. This, however, is only due to the fact that all these observers are on the earth, and share its motion. Even two observers in aeroplanes moving in opposite directions would have at the most a relative velocity of eight thousand miles an hour, which is very little in comparison with 186,000 miles a second (the velocity of light). If an electron with a velocity of 170,000 miles a second could observe the time between the two flashes, it would arrive at a quite different estimate, after making full allowance for the velocity of light. How do you know this? the reader may ask. You are not an electron, you cannot move at these terrific speeds, no man of science has ever made the observations which would prove the truth of your assertion. Nevertheless, as we shall see in the sequel, there is good ground for the assertion—ground, first of all, in experiment, and—what is remarkable—ground in reasonings which could have been made at any time, but were not made until experiments had shown that the old reasonings must be wrong.

There is a general principle to which the theory of relativity appeals, which turns out to be more powerful than anybody would suppose. If you know that one man is twice as rich as another, this fact must appear equally whether you estimate the wealth of both in pounds or dollars or francs or any other currency. The numbers representing their fortunes will be changed, but one number will always be double the other. The same sort of thing, in more complicated forms, reappears in physics. Since all motion is relative, you may take any body you like as your standard body of reference, and estimate all other motions with reference to that one. If you are in a train and walking to the dining-car, you naturally, for the moment, treat the train as fixed and estimate your motion in relation to it. But when you think of the journey you are making, you think of the earth as fixed, and say you are moving at the rate of sixty miles an hour. An astronomer who is concerned with the solar system takes the sun as fixed, and regards you as rotating and revolving; in comparison with this motion, that of the train is so slow that it hardly counts. An astronomer who is interested in the stellar universe may add the motion of the sun relatively to the average of the stars. You cannot say that one of these ways of estimating your motion is more correct than another; each is perfectly correct as soon as the reference-body is assigned. Now just as you can estimate a man's fortune in different currencies without altering its relations to the fortunes of other men, so you can estimate a body's motion by means of different reference bodies without altering its relations to other motions. And as physics is entirely concerned with relations, it must be possible to express all the laws of physics by referring all motions to any given body as the standard.

We may put the matter in another way. Physics is intended to give information about what really occurs in the physical world, and not only about the private perceptions of separate observers. Physics must, therefore, be concerned with those features which a physical process has in common for all observers, since such features alone can be regarded as belonging to the physical occurrence itself. This requires that the *laws* of phenomena should be the same whether the phenomena are described as they appear to one observer or as they appear to another. This single principle is the generating motive of the whole theory of relativity.

Now what we have hitherto regarded as the spatial and

temporal properties of physical occurrences are found to be in large part dependent upon the observer; only a residue can be attributed to the occurrences in themselves, and only this residue can be involved in the formulation of any physical law which is to have an *a priori* chance of being true. Einstein found ready to his hand an instrument of pure mathematics, called the theory of tensors, which enabled him to discover laws expressed in terms of the objective residue and agreeing approximately with the old laws. Where Einstein's laws differed from the old ones, they have hitherto proved more in accord with observation.

If there were no reality in the physical world, but only a number of dreams dreamed by different people, we should not expect to find any laws connecting the dreams of one man with the dreams of another. It is the close connection between the perceptions of one man and the (roughly) simultaneous perceptions of another that makes us believe in a common external origin of the different related perceptions. Physics account both for the likenesses and for the differences between different people's perceptions of what we call the 'same' occurrence. But in order to do this it is first necessary for the physicist to find out just what are the likenesses. They are not quite those traditionally assumed, because neither space nor time separately can be taken as strictly objective. What is objective is a kind of mixture of the two called 'space-time.' To explain this is not easy, but the attempt must be made; it will be begun in the next chapter.

CHAPTER III

The Velocity of Light

MOST of the curious things in the theory of relativity are connected with the velocity of light. If the reader is to grasp the reasons for such a serious theoretical reconstruction, he must have some idea of the facts which made the old system break down.

The fact that light is transmitted with a definite velocity was first established by astronomical observations. Jupiter's moons are sometimes eclipsed by Jupiter, and it is easy to calculate the times when this ought to occur. It was found that when Jupiter was unusually near the earth an eclipse of one of his moons would be observed a few minutes earlier than was expected; and when Jupiter was unusually remote, a few minutes later than was expected. It was found that these deviations could all be accounted for by assuming that light has a certain velocity, so that what we observe to be happening in Jupiter really happened a little while ago—longer ago when Jupiter is distant than when it is near. Just the same velocity of light was found to account for similar facts in regard to other parts of the solar system. It was therefore accepted that light *in vacuo* always travels at a certain constant rate, almost exactly 300,000 kilometres a second. (A kilometre is about five-eighths of a mile.) When it became established that light consists of waves, this velocity was that of propagation of waves in the aether—at least they used to be in the aether, but now the aether has grown somewhat shadowy, though the waves remain. This same velocity is that of radio waves (which are like light-waves, only longer) and of X-rays (which are like light-waves, only shorter). It is generally held nowadays to be the velocity with which gravitation is propagated (before the discovery of relativity theory, it was thought that gravitation was propagated instantaneously, but this view is now untenable).

So far, all is plain sailing. But as it became possible to make more accurate measurements, difficulties began to accumulate. The waves were supposed to be in the aether, and therefore their velocity ought to be relative to the aether. Now since the aether (if it exists) clearly offers no resistance to the motions of the heavenly bodies, it would seem natural to suppose that it does not share their motion. If the earth had to push a lot of aether before it, in the sort of way that a steamer pushes water before it, one would expect a resistance on the part of the aether analogous to that offered by the water to the steamer. Therefore the general view was that the aether could pass through bodies without difficulty, like air through a coarse sieve, only more so. If this were the case, then the earth in its orbit must have a velocity relative to the aether. If, at some one point of its orbit, it happened to be moving exactly with the aether, it must at other points be moving through it all the faster. If you go for a circular walk on a windy day, you must be walking against the wind part of the way, whatever wind may be blowing; the principle in this case is the same. It follows that, if you choose two days six months apart, when the earth in its orbit is moving in exactly opposite directions, it must be moving against an aether-wind on at least one of these days.

Now if there is an aether wind, it is clear that, relatively to an observer on the earth, light-signals will seem to travel faster with the wind than across it, and faster across it than against it. This is what Michelson and Morley set themselves to test by their famous experiment. They sent out light-signals in two directions at right angles; each was reflected from a mirror, and came back to the place from which both had been sent out. Now anybody can verify, either by trial or by a little arithmetic, that it takes longer to row a given distance on a river up-stream and then back again, than it takes to row the same distance across the stream and back again. Therefore, if there were an aether wind, one of the two light-signals, which consist of waves in the aether, ought to have travelled to the mirror and back at a slower average rate than the other. Michelson and Morley tried the experiment, they tried it in various positions, they tried it again later. Their apparatus was quite accurate enough to have detected the expected difference of speed or even a much smaller difference, if it had existed, but not the smallest difference could be observed. The result was a surprise to them as to everybody else; but careful repetitions made doubt impossible. The experiment was first made as long ago as 1881, and was repeated with more elaboration in 1887. But it was many years before it could be rightly interpreted.

The supposition that the earth carries the neighbouring aether with it in its motion was found to be impossible, for a number of reasons. Consequently a logical deadlock seemed to have arisen, from which at first physicists sought to extricate themselves by very arbitrary hypotheses. The most important of these was that of Fitzgerald developed by Lorentz, and known as the Fitzgerald contraction hypothesis.

According to this hypothesis, when a body is in motion it becomes shortened in the direction of motion by a certain proportion depending upon its velocity. The amount of the contraction was to be just enough to account for the negative result of the Michelson–Morley experiment. The journey upstream and down again was to have been really a shorter journey than the one across the stream, and was to have been just so much shorter as would enable the slower light-wave to traverse it in the same time. Of course the shortening could never be detected by measurement, because our measuring rods would share it. A foot-rule placed in the line of the earth's motion would be shorter than the same foot-rule placed at right angles to the earth's motion. This point of view resembles nothing so much as the White Knight's 'plan to dye one's whiskers green, and always use so large a fan that they could not be seen.' The odd thing was that the plan worked well enough. Later on, when Einstein propounded his special theory of relativity (1905), it was found that the hypothesis was in a certain sense correct, but only in a certain sense. That is to say, the supposed contraction is not a physical fact, but a result of certain conventions of measurement which, when once the right point of view has been found, are seen to be such as we are almost compelled to adopt. But I do not wish yet to set forth Einstein's solution to the puzzle. For the present, it is the nature of the puzzle itself that I want to make clear.

On the face of it, and apart from hypotheses *ad hoc*, the Michelson–Morley experiment (in conjunction with others) showed that, relatively to the earth, the velocity of light is the same in all directions, and that this is equally true at all times of the year, although the direction of the earth's motion is always changing as it goes round the sun. Moreover it appeared that this is not a peculiarity of the earth, but is true of all bodies: if a light-signal is sent out from a body, that body will remain at the centre of the waves as they travel outwards, no matter how it may be moving—at least that will be the view of observers moving with the body. This was the plain and natural meaning of the experiments, and Einstein succeeded in inventing a theory which accepted it. But at first it was thought logically impossible to accept this plain and natural meaning.

A few illustrations will make it clear how very odd the facts are. When a shell is fired, it moves faster than sound: the people at whom it is fired first see the flash, then (if they are lucky) see the shell go by, and last of all hear the report. It is clear that if you could put a scientific observer on the shell, he would never hear the report, as the shell would burst and kill him before the sound had overtaken him. But if sound worked on the same principles as light, our observer would hear everything just as if he were at rest. In that case, if a screen, suitable for producing echoes, were attached to the shell and travelling with it, say a hundred yards in front of it, our observer would hear the echo of the report from the screen after just the same interval of time as if he and the shell were at rest. This, of course, is an experiment which cannot be performed, but others which can be performed will show the difference. We might find some place on a railway where there is an echo from a place further along the railway—say a place where the railway goes into a tunnel—and when a train is travelling along the railway, let a man on the bank fire a gun. If the train is travelling towards the echo, the passengers will hear the echo sooner than the man on the bank; if it is travelling in the opposite direction, they will hear it later. But these are not quite the circumstances of the Michelson–Morley experiment. The mirrors in that experiment correspond to the echo, and the mirrors are moving with the earth, so the echo ought to move with the train. Let us suppose that the shot is fired from the guard's-van, and the echo comes from a screen on the engine. We will suppose the distance from the guard's-van to the engine to be the distance that sound can travel in a second (about one-fifth of a mile), and the speed of the train to be one-twelfth of the speed of sound (about sixty miles an hour). We now have an experiment which can be performed by the people in the train. If the train were at rest, the guard would hear the echo in two seconds; as it is, he will hear it in two and $\frac{2}{143}$ seconds. From this difference, if he knows the velocity of sound, he can calculate the velocity of the train, even if it is a foggy night so that he cannot see the banks. But if sound behaved like light, he would hear the echo in two seconds however fast the train might be travelling.

Various other illustrations will help to show how extraordinary—from the point of view of tradition and commonsense—are the facts about the velocity of light. Every one knows that if you are on an escalator you reach the top sooner if you walk up than if you stand still. But if the escalator moved with the velocity of light (which it does not do even in New York), you would reach the top at exactly the same moment whether you walked up or stood still. Again: if you are walking along a road at the rate of four miles an hour, and a motor-car passes you going in the same direction at the rate of forty miles an hour, if you and the motor-car both keep going the distance between you after an hour will be thirty-six miles. But if the motor-car met you, going in the opposite direction, the distance after an hour would be forty-four miles. Now if the motor-car were travelling with the velocity of light, it would make no difference whether it met or passed you: in either case, it would, after a second, be 186,000 miles from you. It would also be 186,000 miles from any other motor-car which happened to be passing or meeting you less rapidly at the previous second. This seems impossible: how can the car be at the same distance from a number of different points along the road?

Let us take another illustration. When a fly touches the surface of a stagnant pool, it causes ripples which move outwards in widening circles. The centre of the circle at any moment is the point of the pool touched by the fly. If the fly moves about over the surface of the pool, it does not remain at the centre of the ripples. But if the ripples were waves of light, and the fly were a skilled physicist, it would find that it always remained at the centre of the ripples, however it might move. Meanwhile a skilled physicist sitting beside the pool would judge, as in the case of ordinary ripples, that the centre was not the fly, but the point of the pool touched by the fly. And if another fly had touched the water at the same spot at the same moment, it also would find that it remained at the centre of the ripples, even if it separated itself widely from the first fly. This is exactly analogous to the Michelson–Morley experiment. The pool corresponds to the aether; the fly corresponds to the earth; the contact of the fly and the pool corresponds to the light-signal which Messrs. Michelson and Morley sent out; and the ripples correspond to the light-waves.

Such a state of affairs seems, at first sight, quite impossible. It is no wonder that, although the Michelson–Morley experiment was made in 1881, it was not rightly interpreted until 1905. Let us see what, exactly, we have been saying. Take the man walking along a road and passed by a motor-car. Suppose there are a number of people at the same point of the road, some walking, some in motor-cars; suppose they are going at varying rates, some in one direction and some in another. I say that if, at this moment, a light-flash is sent out from the place where they all are, the light-waves will be 186,000 miles from each one of them after a second by his watch, although the travellers will not any longer be all in the same place. At the end of a second by your watch it will be 186,000 miles from you, and it will also be 186,000 miles from a person who met you when it was sent out, but was moving in the opposite direction, after a second by his watch—assuming both to be perfect watches. How can this be?

There is only one way of explaining such facts, and that is, to assume that watches and clocks are affected by motion. I do not mean that they are affected in ways that could be remedied by greater accuracy in construction; I mean something much more fundamental. I mean that, if you say an hour has elapsed between two events, and if you base this assertion upon ideally careful measurements with ideally accurate chronometers, another equally precise person, who has been moving rapidly relatively to you, may judge that the time was more or less than an hour. You cannot say that one is right and the other wrong, any more than you could if one used a clock showing Greenwich time and another a clock showing New York time. How this comes about, I shall explain in the next chapter.

There are other curious things about the velocity of light. One is, that no material body can ever travel as fast as light, however great may be the force to which it is exposed, and however long the force may act. An illustration may help to make this clear. At exhibitions one sometimes sees a series of moving platforms, going round and round in a circle. The outside one goes at four miles an hour; the next goes four miles an hour faster than the first; and so on. You can step across from each to the next, until you find yourself going at a tremendous pace. Now you might think that, if the first platform does four miles an hour, and the second does four miles an hour relatively to the first, then the second does eight miles an hour relatively to the ground. This is an error; it does a little less, though so little less that not even the most careful measurements could detect the difference. I want to make quite clear what it is that I mean. I will suppose that, in the morning, when the apparatus is just about to start, three men with ideally accurate chronometers stand in a row, one on the ground, one on the first platform, and one on the second. The first platform moves at the rate of four miles an hour with respect to the ground. Four miles an hour is 352 feet in a minute. The man on the ground, after a minute by his watch, notes the place on the ground opposite the man on the first platform, who has been standing still while the platform carried him along. The man on the ground measures the distance on the ground from himself to the point opposite the man on the first platform, and finds it is 352 feet. The man on the first platform, after a minute by his watch, notes the point on his platform opposite to the man on the second platform. The man on the first platform measures the distance from himself to the point opposite the man on the second platform; it is again 352 feet. Problem: how far will the man on the ground judge that the man on the second platform has travelled in a minute? That is to say, if the man on the ground, after a minute by his watch, notes the place on the ground opposite the man on the second platform, how far will this be from the man on the ground? You would say, twice 352 feet, that is to say, 704 feet. But in fact it will be a little less, though so little less as to be inappreciable. The discrepancy is owing to the fact that the two watches do not keep perfect time, in spite of the fact that each is accurate from its owner's point of view. If you had a long series of such moving platforms, each moving four miles an hour relatively to the one before it, you would never reach a point where the last moving was with the velocity of light relatively to the ground, not even if you had millions of them. The discrepancy, which is very small for small velocities, becomes greater as the velocity increases, and makes the velocity of light an unattainable limit. How all this happens, is the next topic with which we must deal.

CHAPTER IV

Clocks and Foot-rules

UNTIL the advent of the special theory of relativity, no one had thought that there could be any ambiguity in the statement that two events in different places happened at the same time. It might be admitted that, if the places were very far apart, there might be difficulty in finding out for certain whether the events were simultaneous, but every one thought the meaning of the question perfectly definite. It turned out, however, that this was a mistake. Two events in distant places may appear simultaneous to one observer who has taken all due precautions to insure accuracy (and, in particular, has allowed for the velocity of light), while another equally careful observer may judge that the first event preceded the second, and still another may judge that the second preceded the first. This would happen if the three observers were all moving rapidly relatively to each other. It would not be the case that one of them would be right and the other two wrong: they would all be equally right. The time-order of events is in part dependent upon the observer; it is not always and altogether an intrinsic relation between the events themselves. Einstein has shown, not only that this view accounts for the phenomena, but also that it is the one which ought to have resulted from careful reasoning based upon the old data. In actual fact, however, no one noticed the logical basis of the theory of relativity until the odd results of experiment had given a jog to people's reasoning powers.

How should we naturally decide whether two events in different places were simultaneous? One would naturally say: they are simultaneous if they are seen simultaneously by a person who is exactly half-way between them. (There is no difficulty about the simultaneity of two events in the *same* place, such, for example, as seeing a light and hearing a noise.) Suppose two flashes of lightning fall in two different places, say Greenwich Observatory and Kew Observatory. Suppose that St. Paul's is half-way between them, and that the flashes appear simultaneous to an observer on the dome of St. Paul's. In that case, a man at Kew will see the Kew flash first, and a man at Greenwich will see the Greenwich flash first, because of the time taken by light to travel over the intervening distance. But all three, if they are ideally accurate observers, will judge that the two flashes were simultaneous, because they will make the necessary allowance for the time of transmission of the light. (I am assuming a degree of accuracy far beyond human powers.) Thus, so far as observers on the earth are concerned, the definition of simultaneity will work well enough, so long as we are dealing with events on the surface of the earth. It gives results which are consistent with each other, and can be used for terrestrial physics in all problems in which we can ignore the fact that the earth moves.

But our definition is no longer so satisfactory when we have two sets of observers in rapid motion relatively to each other. Suppose we see what would happen if we substitute sound for light, and define two occurrences as simultaneous when they are heard simultaneously by a man half-way between them. This alters nothing in the principle, but makes the matter easier owing to the much slower velocity of sound. Let us suppose that on a foggy night two men belonging to a gang of brigands shoot the guard and engine-driver of a train. The guard is at the end of the train; the brigands are on the line, and shoot their victims at close quarters. An old gentleman who is exactly in the middle of the train hears the two shots simultaneously. You would say, therefore, that the two shots were simultaneous. But a station-master who is exactly half-way between the two brigands hears the shot which kills the guard first. An Australian millionaire uncle of the guard and engine-driver (who are cousins) has left his whole fortune to the guard, or, should he die first, to the engine-driver. Vast sums are involved in the question which died first. The case goes to the House of Lords, and the lawyers on both sides, having been educated at Oxford, are agreed that either the old gentleman or the station-master must have been mistaken.

In fact, both may perfectly well be right. The train travels away from the shot at the guard, and towards the shot at the engine-driver; therefore the noise of the shot at the guard has farther to go before reaching the old gentleman than the shot at the engine-driver has. Therefore if the old gentleman is right in saying that he heard the two reports simultaneously, the station-master must be right in saying that he heard the shot at the guard first.

We, who live on the earth, would naturally, in such a case, prefer the view of simultaneity obtained from a person at rest on the earth to the view of a person travelling in a train. But in theoretical physics no such parochial prejudices are permissible. A physicist on a comet, if there were one, would have just as good a right to his view of simultaneity as an earthly physicist has to his, but the results would differ, in just the same sort of way as in our illustration of the train and the shots. The train is not any more 'real' in motion than the earth; there is no 'really' about it. You might imagine a rabbit and a hippopotamus arguing as to whether man is 'really' a large animal; each would think his own point of view the natural one, and the other a pure flight of fancy. There is just as little substance in an argument as to whether the earth or the train is 'really' in motion. And therefore, when we are defining simultaneity between distant events, we have no right to pick and choose among different bodies to be used in defining the point halfway between the events. All bodies have an equal right to be chosen. But if, for one body, the two events are simultaneous according to the definition, there will be other bodies for which the first precedes the second, and still others for which the second precedes the first. We cannot therefore say unambiguously that two events in distant places are simultaneous. Such a statement only acquires a definite meaning in relation to a definite observer. It belongs to the subjective part of our observation of physical phenomena, not to the objective part which is to enter into physical laws.

This question of time in different places is perhaps, for the imagination, the most difficult aspect of the theory of relativity. We are accustomed to the idea that everything can be dated.

Historians make use of the fact that there was an eclipse of the sun visible in China on August 29, in the year 776 B.C.[1] No doubt astronomers could tell the exact hour and minute when the eclipse began to be total at any given spot in North China. And it seems obvious that we can speak of the positions of the planets at a given instant. The Newtonian theory enables us to calculate the distance between the earth and (say) Jupiter at a given time by the Greenwich clocks; this enables us to know how long light takes at that time to travel from Jupiter to the earth—say half an hour; this enables us to infer that half an hour ago Jupiter was where we see it now. All this seems obvious. But in fact it only works in practice because the relative velocities of the planets are very small compared with the velocity of light. When we judge that an event on the earth and an event on Jupiter have happened at the same time—for example, that Jupiter eclipsed one of his moons when the Greenwich clocks showed twelve midnight—a person moving rapidly relatively to the earth would judge differently, assuming that both he and we had made the proper allowance for the velocity of light. And naturally the disagreement about simultaneity involves a disagreement about periods of time. If we judged that two events on Jupiter were separated by twenty-four hours, another person might judge that they were separated by a longer time, if he were moving rapidly relatively to Jupiter and the earth.

The universal cosmic time which used to be taken for granted is thus no longer admissible. For each body, there is a definite time-order for the events in its neighbourhood; this may be called the 'proper' time for that body. Our own experience is governed by the proper time for our own body. As we all remain very nearly stationary on the earth, the proper times of different human beings agree, and can be lumped together as terrestrial time. But this is only the time appropriate to *large*

[1] A contemporary Chinese ode, after giving the day of the year correctly, proceeds:

> 'For the moon to be eclipsed
> Is but an ordinary matter.
> Now that the sun had been eclipsed
> How bad it is!'

bodies on the earth. For beta particles (electrons) in laboratories, quite different times would be wanted; it is because we insist upon using our own time that these particles seem to increase in mass with rapid motion. From their own point of view, their mass remains constant, and it is we who suddenly grow thin or corpulent. The history of a physicist as observed by a beta particle would resemble Gulliver's travels.

The question now arises: what really is measured by a clock? When we speak of a clock in the theory of relativity, we do not mean only clocks made by human hands: we mean anything which goes through some regular periodic performance. The earth is a clock, because it rotates once in every twenty-three hours and fifty-six minutes. An atom is a clock, because it emits light-waves of very definite frequencies; these are visible as bright lines in the spectrum of the atom. The world is full of periodic occurrences, and fundamental mechanisms, such as atoms, show an extraordinary similarity in different parts of the universe. Any one of these periodic occurrences may be used for measuring time; the only advantage of humanly manufactured clocks is that they are specially easy to observe. However, some of the others are more accurate. Nowadays the short radio waves emitted under certain conditions by caesium atoms and ammonia molecules are being used to establish standards of time measurement more uniform than those based on the earth's rotation. But the question remains: If cosmic time is abandoned, what is really measured by a clock in the wide sense that we have just given to the term?

Each clock gives a correct measure of its own 'proper' time, which, as we shall see presently, is an important physical quantity. But it does not give an accurate measure of any physical quantity connected with events on bodies that are moving rapidly in relation to it. It gives one datum towards the discovery of a physical quantity connected with such events, but another datum is required, and this has to be derived from measurement of distances in space. Distances in space, like periods of time, are in general not objective physical facts, but partly dependent upon the observer. How this comes about must now be explained.

First of all, we have to think of the distance between two events, not between two bodies. This follows at once from what we have found as regards time. If two bodies are moving relatively to each other—and this is really always the case—the distance between them will be continually changing, so that we can only speak of the distance between them at a given time. If you are in a train travelling towards Edinburgh, we can speak of your distance from Edinburgh at a given time. But, as we said, different observers will judge differently as to what is the 'same' time for an event in the train and an event in Edinburgh. This makes the measurement of distances relative, in just the same way as the measurement of times has been found to be relative. We commonly think that there are two separate kinds of interval between two events, an interval in space and an interval in time: between your departure from London and your arrival in Edinburgh, there are four hundred miles and ten hours. We have already seen that another observer will judge the time differently; it is even more obvious that he will judge the distance differently. An observer on the sun will think the motion of the train quite trivial, and will judge that you have travelled the distance travelled by the earth in its orbit and its diurnal rotation. On the other hand, a flea in the railway carriage will judge that you have not moved at all in space, but have afforded him a period of pleasure which he will measure by his 'proper' time, not by Greenwich Observatory. It cannot be said that you or the sun-dweller or the flea are mistaken: each is equally justified and is only wrong if he ascribes an objective validity to his subjective measures. The distance in space between two events is, therefore, not in itself a physical fact. But, as we shall see, there is a physical fact which can be inferred from the distance in time together with the distance in space. This is what is called the 'interval' in space-time.

Taking any two events in the universe, there are two different possibilities as to the relation between them. It may be physically possible for a body to travel so as to be present at both events or it may not. This depends upon the fact that no body can travel as fast as light. Suppose, for example, that it were possible to send out a flash of light from the earth and have it reflected back from the moon. (This is an experiment which has actually been performed, with laser light, and with radar waves, which travel at the same speed.) The time between the sending of the flash and the return of the reflection would be about two and a half seconds. No body could travel so fast as to be present on the earth during any part of those two and a half seconds and also present on the moon at the moment of the arrival of the flash, because in order to do so the body would have to travel faster than light. But theoretically a body could be present on the earth at any time before or after those two and a half seconds and also present on the moon at the time when the flash arrived. When it is physically impossible for a body to travel so as to be present at both events, we shall say that the interval[1] between the two events is 'space-like'; when it is physically possible for a body to be present at both events, we shall say that the interval between the two events is 'time-like.' When the interval is 'space-like,' it is possible for a body to move in such a way that an observer on the body will judge the two events to be simultaneous. In that case, the 'interval' between the two events is what such an observer will judge to be the distance in space between them. When the interval is 'time-like,' a body can be present at both events; in that case, the 'interval' between the two events is what an observer on the body will judge to be the time between them, that is to say, it is his 'proper' time between the two events. There is a limiting case between the two, when the two events are parts of one light-flash—or, as we might say, when the one event is the seeing of the other. In that case, the interval between the two events is zero.

There are thus three cases. (1) It may be possible for a ray of light to be present at both events; this happens whenever one of them is the seeing of the other. In this case the interval between the two events is zero. (2) It may happen that no body can travel from one event to the other, because in order to do so it would have to travel faster than light. In that case, it is always physically possible for a body to travel in such a

[1] I shall define 'interval' in a moment.

way that an observer on the body would judge the two events to be simultaneous. The interval is what he would judge to be the distance in space between the two events. Such an interval is called 'space-like.' (3) It may be physically possible for a body to travel so as to be present at both events; in that case, the interval between them is what an observer on such a body will judge to be the time between them. Such an interval is called 'time-like.'

The interval between two events is a physical fact about them, not dependent upon the particular circumstances of the observer.

There are two forms of the theory of relativity, the special and the general. The former is in general only approximate, but becomes very nearly exact at great distances from gravitating matter. Whenever gravitation may be neglected, the special theory can be applied, and then the interval between two events can be calculated when we know the distance in space and the distance in time between them, estimated by any observer. If the distance in space is greater than the distance that light would have travelled in the time, the separation is space-like. Then the following construction gives the interval between the two events: Draw a line AB as long as the distance that light would travel in the time; round A describe a circle whose radius is the distance in space between the two events; through B draw BC perpendicular to AB, meeting the circle in C. Then BC is the length of the interval between the two events.

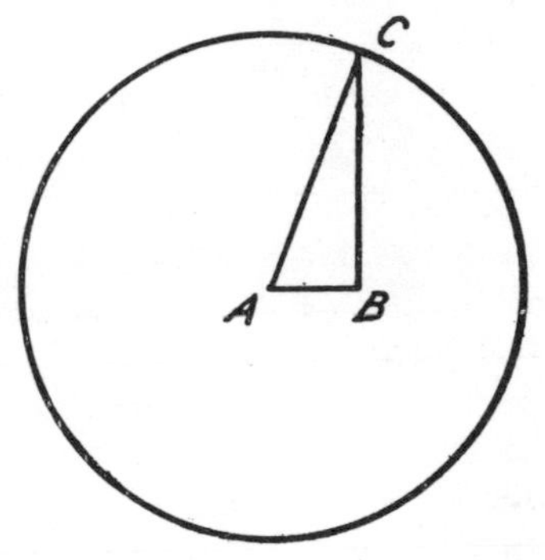

When the distance is time-like, use the same figure, but let AC be now the distance that light would travel in the time, while AB is the distance in space between the two events. The interval between them is now the time that light would take to travel the distance BC.

Although AB and AC are different for different observers,

BC is the same length for all observers, subject to corrections made by the general theory. It represents the one interval in 'space-time' which replaces the two intervals in space and time of the older physics. So far, this notion of interval may appear somewhat mysterious, but as we proceed it will grow less so, and its reason in the nature of things will gradually emerge.

CHAPTER V

Space-Time

EVERYBODY who has ever heard of relativity knows the phrase 'space-time,' and knows that the correct thing is to use this phrase when formerly we should have said 'space *and* time.' But very few people who are not mathematicians have any clear idea of what is meant by this change of phraseology. Before dealing further with the special theory of relativity, I want to try to convey to the reader what is involved in the new phrase 'space-time,' because that is, from a philosophical and imaginative point of view, perhaps the most important of all the novelties that Einstein introduced.

Suppose you wish to say where and when some event has occurred—say an explosion on an airship—you will have to mention four quantities, say the latitude and longitude, the height above the ground, and the time. According to the traditional view, the first three of these give the position in space, while the fourth gives the position in time. The three quantities that give the position in space may be assigned in all sorts of ways. You might, for instance, take the plane of the equator, the plane of the meridian of Greenwich, and the plane of the 90th meridian, and say how far the airship was from each of these planes; these three distances would be what are called 'Cartesian co-ordinates,' after Descartes. You might take any other three planes all at right angles to each other, and you would still have Cartesian co-ordinates. Or you might take the distance from London to a point vertically below the airship, the direction of this distance (north-east, west-south-west, or whatever it might be), and the height of the airship above the ground. There are an infinite number of such ways of fixing the position in space, all equally legitimate; the choice between them is merely one of convenience.

When people said that space had three dimensions, they

meant just this: that three quantities were necessary in order to specify the position of a point in space, but that the method of assigning these quantities was wholly arbitrary.

With regard to time, the matter was thought to be quite different. The only arbitrary elements in the reckoning of time were the unit, and the point of time from which the reckoning started. One could reckon in Greenwich time, or in Paris time, or in New York time; that made a difference as to the point of departure. One could reckon in seconds, minutes, hours, days, or years; that was a difference of unit. Both these were obvious and trivial matters. There was nothing corresponding to the liberty of choice as to the method of fixing position in space. And, in particular, it was thought that the method of fixing position in space and the method of fixing position in time could be made wholly independent of each other. For these reasons, people regarded time and space as quite distinct.

The theory of relativity has changed this. There are now a number of different ways of fixing position in time, which do not differ merely as to the unit and the starting-point. Indeed, as we have seen, if one event is simultaneous with another in one reckoning, it will precede it in another, and follow it in a third. Moreover, the space and time reckonings are no longer independent of each other. If you alter the way of reckoning position in space, you may also alter the time-interval between two events. If you alter the way of reckoning time, you may also alter the distance in space between two events. Thus space and time are no longer independent, any more than the three dimensions of space are. We still need four quantities to determine the position of an event, but we cannot, as before, divide off one of the four as quite independent of the other three.

It is not quite true to say that there is no longer any distinction between time and space. As we have seen, there are time-like intervals and space-like intervals. But the distinction is of a different sort from that which was formerly assumed. There is no longer a universal time which can be applied without ambiguity to any part of the universe; there are only the various 'proper' times of the various bodies in the universe, which agree approximately for two bodies which are not in rapid motion, but never agree exactly except for two bodies which are at rest relatively to each other.

The picture of the world which is required for this new state of affairs is as follows: Suppose an event E occurs to me, and simultaneously a flash of light goes out from me in all directions. Anything that happens to any body after the light from the flash has reached it is definitely after the event E in any system of reckoning time. Any event anywhere which I could have seen before the event E occurred to me is definitely before the event E in any system of reckoning time. But any event which happened in the intervening time is not definitely either before or after the event E. To make the matter definite: suppose I could observe a person in Sirius, and he could observe me. Anything which he does, and which I see before the event E occurs to me, is definitely before E; anything he does after he has seen the event E is definitely after E. But anything that he does before he sees the event E, but so that I see it after the event E has happened, is not definitely before or after E. Since light takes many years to travel from Sirius to the earth, this gives a period of twice as many years in Sirius which may be called 'contemporary' with E, since these years are not definitely before or after E.

Dr. A. A. Robb, in his *Theory of Time and Space*, suggested a point of view which may or may not be philosophically fundamental, but is at any rate a help in understanding the state of affairs we have been describing. He maintained that one event can only be said to be definitely *before* another if it can influence that other in some way. Now influences spread from a centre at varying rates. Newspapers exercise an influence emanating from London at an average rate of about twenty miles an hour—rather more for long distances. Anything a man does because of what he reads in the newspaper is clearly subsequent to the printing of the newspaper. Sounds travel much faster: it would be possible to arrange a series of loud-speakers along the main roads, and have newspapers shouted from each to the next. But telegraphing is quicker, and wireless telegraphy travels with the velocity of light, so that nothing quicker can ever be hoped for. Now what a man does in consequence of receiving a wireless message he does *after* the message was sent; the meaning here is quite independent of conventions as to the measurement of time. But anything that he does while the message is on its way cannot be influenced by the sending of the message, and cannot influence the sender until some little time after he sent the message, that is to say, if two bodies are widely separated, neither can influence the other except after a certain lapse of time; what happens before that time has elapsed cannot affect the distant body. Suppose, for instance, that some notable event happens on the sun: there is a period of sixteen minutes on the earth during which no event on the earth can have influenced or been influenced by the said notable event on the sun. This gives a substantial ground for regarding that period of sixteen minutes on the earth as neither before nor after the event on the sun.

The paradoxes of the special theory of relativity are only paradoxes because we are unaccustomed to the point of view, and in the habit of taking things for granted when we have no right to do so. This is especially true as regards the measurement of lengths. In daily life, our way of measuring lengths is to apply a foot-rule or some other measure. At the moment when the foot-rule is applied, it is at rest relatively to the body which is being measured. Consequently the length that we arrive at by measurement is the 'proper' length, that is to say, the length as estimated by an observer who shares the motion of the body. We never, in ordinary life, have to tackle the problem of measuring a body which is in continual motion. And even if we did, the velocities of visible bodies on the earth are so small relatively to the earth that the anomalies dealt with by the theory of relativity would not appear. But in astromony, or in the investigation of atomic structure, we are faced with problems which cannot be tackled in this way. Not being Joshua, we cannot make the sun stand still while we measure it; if we are to estimate its size we must do so while it is in motion relatively to us. And similarly if you want to estimate the size of an electron, you have to do so while it is in rapid motion, because it never stands still for a moment. This is the sort of problem with which the theory of relativity is concerned.

Measurement with a foot-rule, when it is possible, gives always the same result, because it gives the 'proper' length of a body. But when this method is not possible, we find that curious things happen, particularly if the body to be measured is moving very fast relatively to the observer. A figure like the one at the end of the previous chapter will help us to understand the state of affairs.

Let us suppose that the body on which we wish to measure lengths is moving relatively to ourselves, and that in one second it moves the distance OM. Let us draw a circle round O whose radius is the distance that light travels in a second. Through M draw MP perpendicular to OM, meeting the circle in P. Thus OP is the distance that light travels in a second. The ratio of OP to OM is the ratio of the velocity of light to the velocity of the body. The ratio of OP to MP is the ratio in which apparent lengths are altered by the motion. That is to say, if the observer judges that two points in the line of motion on the moving body are at a distance from each other represented by MP, a person moving with the body would judge that they were at a distance represented (on the same scale) by OP. Distances on the moving body at right angles to the line of motion are not affected by the motion. The whole thing is reciprocal; that is to say, if an observer moving with the body were to measure lengths on the previous observer's body, they would be altered in just the same proportion. When two bodies are moving relatively to each other, lengths on either appear shorter to the other than to themselves. This is the Fitzgerald contraction, which was first invented to account for the result of the Michelson–Morley experiment. But it now emerges naturally from the fact that the two observers do not make the same judgment of simultaneity.

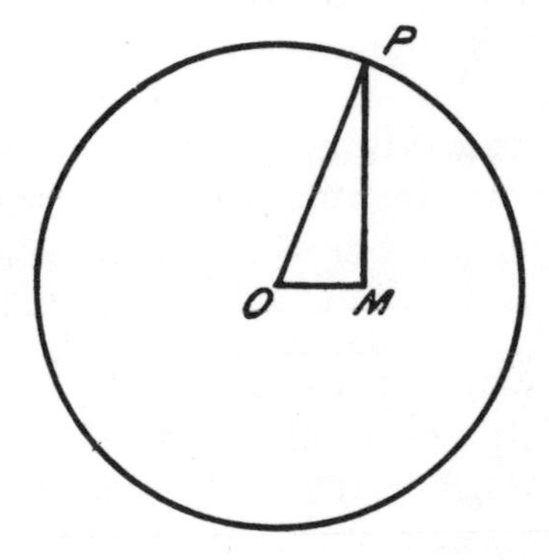

The way in which simultaneity comes in is this: We say that

two points on a body are a foot apart when we can *simultaneously* apply one end of a foot-rule to the one and the other end to the other. If, now, two people disagree about simultaneity, and the body is in motion, they will obviously get different results from their measurements. Thus the trouble about time is at the bottom of the trouble about distance.

The ratio of OP to MP is the essential thing in all these matters. Times and lengths and masses are all altered in this proportion when the body concerned is in motion relatively to the observer. It will be seen that, if OM is very much smaller than OP, that is to say, if the body is moving very much more slowly than light, MP and OP are very nearly equal, so that the alterations produced by the motion are very small. But if OM is nearly as large as OP, that is to say, if the body is moving nearly as fast as light, MP becomes very small compared to OP, and the effects become very great. The apparent increase of mass in swiftly moving particles had been observed, and the right formula had been found, before Einstein invented his special theory of relativity. In fact, Lorentz had arrived at the formulae called the 'Lorentz transformation,' which embody the whole mathematical essence of the special theory of relativity. But it was Einstein who showed that the whole thing was what we ought to have expected, and not a set of makeshift devices to account for surprising experimental results. Nevertheless, it must not be forgotten that experimental results were the original motive of the whole theory, and have remained the ground for undertaking the tremendous logical reconstruction involved in Einstein's theories.

We may now recapitulate the reasons which have made it necessary to substitute 'space-time' for space and time. The old separation of space and time rested upon the belief that there was no ambiguity in saying that two events in distant places happened at the same time; consequently it was thought that we could describe the topography of the universe at a given instant in purely spatial terms. But now that simultaneity has become relative to a particular observer, this is no longer possible. What is, for one observer, a description of the state of the world at a given instant, is, for another observer, a series of events at various different times, whose relations are not merely spatial but also temporal. For the same reason, we are concerned with *events*, rather than with *bodies*. In the old theory, it was possible to consider a number of bodies all at the same instant, and since the time was the same for all of them it could be ignored. But now we cannot do that if we are to obtain an objective account of physical occurrences. We must mention the date at which a body is to be considered, and thus we arrive at an 'event,' that is to say, something which happens at a given time. When we know the time and place of an event in one observer's system of reckoning, we can calculate its time and place according to another observer. But we must know the time as well as the place, because we can no longer ask what is its place for the new observer at the 'same' time as for the old observer. There is no such thing as the 'same' time for different observers, unless they are at rest relatively to each other. We need four measurements to fix a position, and four measurements fix the position of an event in space-time, not merely of a body in space. Three measurements are not enough to fix any position. That is the essence of what is meant by the substitution of space-time for space and time.

CHAPTER VI

The Special Theory of Relativity

THE special theory of relativity arose as a way of accounting for the facts of electro-magnetism. We have here a somewhat curious history. In the eighteenth and early nineteenth centuries, the theory of electricity was wholly dominated by the Newtonian analogy. Two electric charges attract each other if they are of different kinds, one positive and one negative, but repel each other if they are of the same kind; in each case, the force varies as the inverse square of the distance, as in the case of gravitation. This force was conceived as an action at a distance, until Faraday, by a number of remarkable experiments, demonstrated the effect of the intervening medium. Faraday was no mathematician; Clerk Maxwell first gave a mathematical form to the results suggested by Faraday's experiments. Moreover Clerk Maxwell gave grounds for thinking that light is an electromagnetic phenomenon, consisting of electromagnetic waves. The medium for the transmission of electromagnetic effects could therefore be taken to be the aether, which had long been assumed for the transmission of light. The correctness of Maxwell's theory of light was proved by the experiments of Hertz in manufacturing electromagnetic waves; these experiments afford the basis for wireless telegraphy. So far, we have a record of triumphant progress, in which theory and experiment alternately assume the leading rôle. At the time of Hertz's experiments, the aether seemed securely established, and in just as strong a position as any other scientific hypothesis not capable of direct verification. But a new set of facts began to be discovered, and gradually the whole picture was changed.

The movement which culminated with Hertz was a movement for making everything continuous. The aether was continuous, the waves in it were continuous, and it was hoped that matter would be found to consist of some continuous structure in the aether. But then came the discovery of the atomic structure of matter, and of the discrete structure of the atoms themselves. Atoms were believed to be built up of electrons, protons, and neutrons. The electron is a small particle bearing a definite charge of negative electricity. The proton bears a definite charge of positive electricity, while the neutron is not charged. (It is only a matter of custom that the charge on the electron is called negative and the charge on the proton positive, rather than the other way round.) It appeared probable that electricity was not to be found except in the form of the charges on the electron and proton; all electrons have exactly the same negative charge, and all protons have an exactly equal and opposite positive charge. Later on other sub-atomic particles were discovered; most of them are called mesons or hyperons. All protons have exactly the same weight; they are about eighteen hundred times as heavy as electrons. All neutrons also have exactly the same weight; they are slightly heavier than protons. Mesons, of which there are several different kinds, weigh more than electrons but less than protons, while hyperons are heavier than protons or neutrons.

Some of the particles bear electric charges, while others do not. It is found that all the positively charged ones have exactly the same charge as the proton, while all the negatively charged ones have exactly the same charge as the electron, although their other properties are quite different. To confuse matters, there is a particle which is identical with the electron, except that it has a positive charge instead of a negative one; it is called the positron. Quite recently a particle has been discovered which is identical with the proton except that it has a negative charge; it is called the anti-proton.

These discoveries about the discrete structure of matter are inseparable from the discoveries of other so-called quantum phenomena, like the bright lines in the spectrum of an atom. It seems that all natural processes show a fundamental discontinuity whenever they can be measured with sufficient precision.

Thus physics has had to digest new facts and face new problems. Although the quantum theory has existed in more or less its present form for forty years, and the special theory of relativity for sixty, little progress was made until about twenty years ago, in connecting the two together. Recent developments in the quantum theory have made it more consistent with relativity, and these improvements have helped our understanding of the sub-atomic particles a good deal, but many serious difficulties remain.

The problems solved by the special theory of relativity in its own right, quite apart from the quantum theory, are typified by the Michelson–Morley experiment. Assuming the correctness of Maxwell's theory of electromagnetism there should have been certain discoverable effects of motion through the aether; in fact, there were none. Then there was the observed fact that a body in very rapid motion appears to increase its mass; the increase is in the ratio of OP to MP in the figure in the preceding chapter. Facts of this sort gradually accumulated until it became imperative to find some theory which would account for them all.

Maxwell's theory reduced itself to certain equations, known as 'Maxwell's equations.' Through all the revolutions which physics has undergone in the last century, these equations have remained standing; indeed they have continually grown in importance as well as in certainty—for Maxwell's arguments in their favour were so shaky that the correctness of his results must almost be ascribed to intuition. Now these equations were, of course, founded upon experiments in terrestrial laboratories, but there was a tacit assumption that the motion of the earth through the aether could be ignored. In certain cases, such as the Michelson–Morley experiment, this ought not to have been possible without measurable error; but it turned out to be always possible. Physicists were faced with the odd difficulty that Maxwell's equations were more accurate than they should be. A very similar difficulty was explained by Galileo at the very beginning of modern physics. Most people think that if you let a weight drop it will fall vertically. But if you try the experiment in the cabin of a moving ship, the weight falls, in relation to the cabin, just as if the ship were at rest; for instance, if it starts from the middle of the ceiling it will drop on to the middle of the floor. That is to say, from the point of view of an observer on the shore it does not fall vertically, since it shares the motion of the ship. So long as the ship's motion is steady, everything goes on inside the ship as if the ship were not moving. Galileo explained how this happens, to the great indignation of the disciples of Aristotle. In orthodox physics, which is derived from Galileo, a uniform motion in a straight line has no discoverable effects. This was, in its day, as astonishing a form of relativity as that of Einstein is to us. Einstein, in the special theory of relativity, set to work to show how electromagnetic phenomena could be unaffected by uniform motion through the aether—if there be an aether. This was a more difficult problem, which could not be solved by merely adhering to the principles of Galileo.

The really difficult effort required for solving this problem was in regard to time. It was necessary to introduce the notion of 'proper' time which we have already considered, and to abandon the old belief in one universal time. The quantitative laws of electromagnetic phenomena are expressed in Maxwell's equation's and these equations are found to be true for any observer, however he may be moving. It is a straightforward mathematical problem to find out what differences there must be between the measures applied by one observer and the measures applied by another, if, in spite of their relative motion, they are to find the same equations verified. The answer is contained in the 'Lorentz transformation,' found as a formula by Lorentz, but interpreted and made intelligible by Einstein.

The Lorentz transformation tells us what estimate of distances and periods of time will be made by an observer whose relative motion is known, when we are given those of another observer. We may suppose that you are in a train on a railway which travels due east. You have been travelling for a time which, by the clocks at the station from which you started, is t. At a distance x from your starting-point, as measured by the people on the line, an event occurs at this moment—say the line is struck by lightning. You have been travelling all the time with a uniform velocity v. The question is: How far from you will you judge that this event has taken place, and how long after you started will it be by your watch, assuming that your watch is correct from the point of view of an observer on the train?

Our solution of this problem has to satisfy certain conditions. It has to bring out the result that the velocity of light is the same for all observers, however they may be moving. And it has to make physical phenomena—in particular, those of electromagnetism—obey the same laws for different observers, however they may find their measures of distances and times affected by their motion. And it has to make all such effects on measurement reciprocal. That is to say, if you are in a train and your motion affects your estimate of distances outside the train, there must be an exactly similar change in the estimate which people outside the train make of distances inside it. These conditions are sufficient to determine the solution of the problem, but the solution requires more mathematics than I have allowed myself in this book.

Before dealing with the matter in general terms, let us take an example. Let us suppose that you are in a train on a long straight railway, and that you are travelling due east at three-fifths of the velocity of light. Suppose that you measure the length of your train, and find that it is a hundred yards. Suppose that the people who catch a glimpse of you as you pass succeed, by skilful scientific methods, in taking observations which enable them to calculate the length of your train. If they do their work correctly, they will find that it is eighty yards long. Everything in the train will seem to them shorter in the direction of the train than it does to you. Dinner plates, which you see as ordinary circular plates, will look to the outsider as if they were oval: they will seem only four-fifths as broad in the direction in which the train is moving as in the direction of the breadth of the train. And all this is reciprocal. Suppose you see out of the window a man carrying a fishing-rod which, by his measurement, is fifteen feet long. If he is holding it upright, you will see it as he does; so you will if he is holding it horizontally at right-angles to the railway. But if he is pointing it along the railway, it will seem to you to be only twelve feet long. In describing what is seen, I have assumed that everyone makes due allowance for perspective. Despite this, all the lengths of objects in the train will be diminished by twenty per cent, in the direction of motion, for people outside, and so will those of objects outside, for you in the train.

But the effects in regard to time are even more strange. This matter has been explained with almost ideal lucidity by Eddington in *Space, Time and Gravitation*. He supposes an aviator travelling, relatively to the earth, at a speed of 161,000 miles a second, and he says:

'If we observed the aviator carefully we should infer that he was unusually slow in his movements; and events in the conveyance moving with him would be similarly retarded—as though time had forgotten to go on. His cigar lasts twice as long as one of ours. I said "infer" deliberately; we should *see* a still more extravagant slowing down of time; but that is easily explained, because the aviator is rapidly increasing his distance from us and the light-impressions take longer and longer to reach us. The more moderate retardation referred to remains after we have allowed for the time of transmission of light. But here again reciprocity comes in, because in the aviator's opinion it is we who are travelling at 161,000 miles a second past him; and when he has made all allowances, he finds that it is we who are sluggish. Our cigar lasts twice as long as his.'

What a situation for envy! Each man thinks that the other's cigar lasts twice as long as his own. It may, however, be some consolation to reflect that the other man's visits to the dentist also last twice as long.

This question of time is rather intricate, owing to the fact that events which one man judges to be simultaneous another considers to be separated by a lapse of time. In order to try to make clear how time is affected, I shall revert to our railway train travelling due east at a rate of three-fifths that of light. For the sake of illustration, I assume that the earth is large and flat, instead of small and round.

If we take events which happen at a fixed point on the earth, and ask ourselves how long after the beginning of the journey they will seem to be to the traveller, the answer is that there will be that retardation that Eddington speaks of, which means in this case that what seems an hour in the life of the stationary person is judged to be an hour and a quarter by the man who observes him from the train. Reciprocally, what seems an hour in the life of the person in the train is judged by the man observing him from outside to be an hour and a quarter. Each makes periods of time observed in the life of the other a quarter as long again as they are to the person who lives through them. The proportion is the same in regard to times as in regard to lengths.

But when, instead of comparing events at the same place on the earth, we compare events at widely separated places, the results are still more odd. Let us now take all the events along the railway, which from the point of view of a person who is stationary on the earth, happen at a given instant, say the instant when the observer in the train passes the stationary person. Of these events, those which occur at points towards which the train is moving will seem to the traveller to have already happened, while those which occur at points behind the train, will for him, be still in the future. When I say that events in the forward direction will seem to have already happened, I am saying something not strictly accurate; because he will not yet have seen them; but when he does see them, he will, after allowing for the velocity of light, come to the conclusion that they must have happened before the moment in question. An event which happens in the forward direction along the railway, and which the stationary observer judges to be now (or rather, will judge to have been now when he comes to know of it) if it occurs at a distance along the line which light could travel in a second, will be judged by the traveller to have occurred three-quarters of a second ago. If it occurs at a distance from the two observers which the man on the earth judges that light could travel in a year, the traveller will judge (when he comes to know of it) that it occurred nine months earlier than the moment when he passes the earth-dweller. And generally, he will ante-date events in the forward direction along the railway by three-quarters of the time that it would take light to travel from them to the man on the earth whom he is just passing, and who holds that these events are happening now—or rather, will hold that they happened now when the light from them reaches him. Events happening on the railway behind the train will be post-dated by an exactly equal amount.

We have thus a two-fold correction to make in the date of an event when we pass from the terrestrial observer to the traveller. We must first take five-fourths of the time as estimated by the earth-dweller, and then subtract three-fourths of the time that it would take light to travel from the event in question to the earth-dweller.

Take some event in a distant part of the universe, which becomes visible to the earth-dweller and the traveller just as they pass each other. The earth-dweller, if he knows how far off the event occurred, can judge how long ago it occurred, since he knows the speed of light. If the event occurred in the direction towards which the traveller is moving, the traveller will infer that it happened twice as long ago as the earth-dweller thinks. But if it occurred in the direction from which he has come, he will argue that it happened only half as long ago as the earth-dweller thinks. If the traveller moves at a different speed, these proportions will be different.

Suppose now that (as sometimes occurs) two new stars have suddenly flared up, and have just become visible to the traveller and to the earth-dweller whom he is passing. Let one of them be in the direction towards which the train is travelling, the other in the direction from which it has come. Suppose that the earth-dweller is able, in some way, to estimate the distance of the two stars, and to infer that light takes fifty years to reach him from the one in the direction towards which the traveller is moving, and one hundred years to reach him from the other. He will then argue that the explosion which produced the new star in the forward direction occurred fifty years ago, while the explosion which produced the other new star occurred a hundred years ago. The traveller will exactly reverse these figures: he will infer that the forward explosion occurred a hundred years ago, and the backward one fifty years ago. I assume that both argue correctly on correct physical data. In fact, both are right, unless they imagine that the other must be wrong. It should be noted that both will have the same estimate of the velocity of light, because their estimates of the distances of the two new stars will vary in exactly the same proportion as their estimates of the times since the explosions. Indeed one of the main motives of this whole theory is to secure that the velocity of light shall be the same for all observers, however they may be moving. This fact, established by experiment, was incompatible with the old theories, and made it absolutely necessary to admit something startling. The theory of relativity is just as little startling as is compatible with the facts. Indeed, after a time, it ceases to seem startling at all.

There is another feature of very great importance in the theory we have been considering, and that is that, although distances and times vary for different observers, we can derive from them the quantity called 'interval,' which is the same for all observers. The 'interval,' in the special theory of relativity, is obtained as follows: take the square of the distance between two events, and the square of the distance travelled by light in the time between the two events; subtract the lesser of these from the greater and the result is defined as the square of the interval between the events. The interval is the same for all observers and represents a genuine physical relation between the two events, which the time and the distance do not. We have already given a geometrical construction for the interval at the end of Chapter IV; this gives the same result as the above rule. The interval is 'time-like' when the time between the events is longer than light would take to travel from the place of the one to the place of the other; in the contrary case it is 'space-like.' When the time between the two events is exactly equal to the time taken by light to travel from one to the other, the interval is zero; the two events are then situated on parts of one light-ray, unless no light happens to be passing that way.

When we come to the general theory of relativity, we shall have to generalize the notion of interval. The more deeply we penetrate into the structure of the world, the more important this concept becomes; we are tempted to say that it is the reality of which distances and period of time are confused representations. The theory of relativity has altered our view of the fundamental structure of the world; that is the source both of its difficulty and of its importance.

The remainder of this chapter may be omitted by readers who have not even the most elementary acquaintance with geometry or algebra. But for the benefit of those whose education has not been *entirely* neglected, I will add a few explanations of the general formula of which I have hitherto given only particular examples. The general formula in question is the 'Lorentz transformation,' which tells, when one body is moving in a given manner relatively to another, how to infer the measures of lengths and times appropriate to the one body from those appropriate to the other. Before giving the algebraical formulae, I will give a geometrical construction. As before, we will suppose that there are two observers, whom we will call O and O′, one of whom is stationary on the earth while the other is travelling at a uniform speed along a straight railway. At the beginning of the time considered, the two observers were at the same point of the railway, but now they are separated by a certain distance. A flash of lightning strikes a point X on the railway, and O judges that at the moment when the flash takes place the observer in the train has reached the point O′. The problem is: how far will O′ judge that he is from the flash, and how long after the beginning of the journey (when he was at O) will he judge that the flash took place? We are supposed to know O's estimates, and we want to calculate those of O′.

In the time that, according to O, has elapsed since the beginning of the journey, let OC be the distance that light would have travelled along the railway. Describe a circle about O, with OC as radius, and through O′ draw a perpendicular to the railway, meeting the circle in D. On OD take a point Y such that OY is equal to OX (X is the point of the railway where the lightning strikes). Draw YM perpendicular to the railway, and OS perpendicular to OD. Let YM and OS meet in S. Also let DO′ produced and OS produced meet in R. Through X and C draw perpendiculars to the railway meeting OS produced in Q and Z respectively. Then RQ (as measured by O) is the distance at which O′ will believe himself to be from the flash, not O′X as it would be according to the old view. And whereas O thinks that, in the time from the beginning of the journey to the flash, light would travel a distance OC, O′ thinks that the time elapsed is that required for light to travel the distance SZ (as measured by O). The interval as measured by O is got by subtracting the square on OX from the square on OC; the interval as measured by O′ is got by subtracting the square

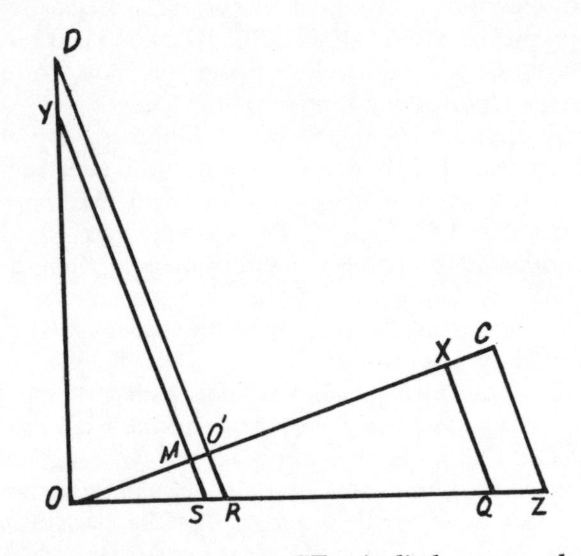

on RQ from the square on SZ. A little very elementary geometry shows that these are equal.

The algebraical formulae embodied in the above construction are as follows: from the point of view of O, let an event occur at a distance x along the railway, and at a time t after the beginning of the journey (when O′ was at O). From the point of view of O′ let the same event occur at a distance x' along the railway, and at a time t' after the beginning of the journey. Let c be the velocity of light, and v the velocity of O′ relative to O. Put

$$\beta = \frac{c}{\sqrt{c^2 - v^2}}$$

Then

$$x' = \beta\,(x - v\,t)$$

$$t' = \beta\left(t - \frac{vx}{c^2}\right)$$

This is the Lorentz transformation, from which everything in this chapter can be deduced.

CHAPTER VII

Intervals in Space-Time

THE special theory of relativity, which we have been considering hitherto, solved completely a certain definite problem: to account for the experimental fact that, when two bodies are in uniform relative motion, all the laws of physics, both those of ordinary dynamics and those connected with electricity and magnetism, are exactly the same for the two bodies. 'Uniform' motion, here, means motion in a straight line with constant velocity. But although one problem was solved by the special theory, another was immediately suggested: what if the motion of the two bodies is not uniform? Suppose, for instance, that one is the earth while the other is a falling stone. The stone has an accelerated motion: it is continually falling faster and faster. Nothing in the special theory enables us to say that the laws of physical phenomena will be the same for an observer on the stone as for one on the earth. This is particularly awkward, as the earth itself is, in an extended sense, a falling body: it has at every moment an acceleration[1] towards the sun, which makes it go round the sun instead of moving in a straight line. As our knowledge of physics is derived from experiments on the earth, we cannot rest satisfied with a theory in which the observer is supposed to have no acceleration. The general theory of relativity removes this restriction, and allows the observer to be moving in any way, straight or crooked, uniformly or with an acceleration. In the course of removing the restriction, Einstein was led to his new law of gravitation, which we shall consider presently. The work was extraordinarily difficult, and occupied him for ten years. The special theory dates from 1905, the general theory from 1915.

[1] Not only an increase in speed, but any change in speed or direction, is called 'acceleration'. The only sort of motion called 'unaccelerated' is motion with constant speed *in a straight line*.

It is obvious from experiences with which we are all familiar that an accelerated motion is much more difficult to deal with than a uniform one. When you are in a train which is travelling steadily, the motion is not noticeable so long as you do not look out of the window; but when the brakes are applied suddenly you are precipitated forwards, and you become aware that something is happening without having to notice anything outside the train. Similarly in a lift everything seems ordinary while it is moving steadily, but at starting and stopping, when its motion is accelerated, you have odd sensations in the pit of the stomach. (We call a motion 'accelerated' when it is getting slower as well as when it is getting quicker; when it is getting slower the acceleration is negative.) The same thing applies to dropping a weight in the cabin of a ship. So long as the ship is moving uniformly, the weight will behave, relatively to the cabin, just as if the ship were at rest: if it starts from the middle of the ceiling, it will hit the middle of the floor. But if there is an acceleration everything is changed. If the boat is increasing its speed very rapidly, the weight will seem to an observer in the cabin to fall in a curve directed towards the stern; if the speed is being rapidly diminished, the curve will be directed towards the bow. All these facts are familiar, and they led Galileo and Newton to regard an accelerated motion as something radically different, in its own nature, from a uniform motion. But this distinction could only be maintained by regarding motion as absolute, not relative. If all motion is relative, the earth is accelerated relatively to the lift just as truly as the lift relatively to the earth. Yet the people on the ground have no sensations in the pits of their stomachs when the lift starts to go up. This illustrates the difficulty of our problem. In fact, though few physicists in modern times have believed in absolute motion, the technique of mathematical physics still embodied Newton's belief in it, and a revolution in method was required to obtain a technique free from this assumption. This revolution was accomplished in Einstein's general theory of relativity.

It is somewhat optional where we begin in explaining the new ideas which Einstein introduced, but perhaps we shall do best by taking the conception of 'interval.' This conception, as it appears in the special theory of relativity, is already a generalization of the traditional notion of distance in space and time; but it is necessary to generalize it still further. However, it is necessary first to explain a certain amount of history, and for this purpose we must go back as far as Pythagoras.

Pythagoras, like many of the greatest characters in history, perhaps never existed: he is a semi-mythical character, who combined mathematics and priestcraft in uncertain proportions. I shall, however, assume that he existed, and that he discovered the theorem attributed to him. He was roughly a contemporary of Confucius and Buddha; he founded a religious sect, which thought it wicked to eat beans, and a school of mathematicians who took a particular interest in right-angled triangles. The theorem of Pythagoras (the 47th proposition of Euclid) states that the sum of the squares on the two shorter sides of a right-angled triangle is equal to the square on the side opposite the right angle. No proposition in the whole of mathematics has had such a distinguished history. We all learned to 'prove' it in youth. It is true that the 'proof' proved nothing, and that the only way to prove it is by experiment. It is also the case that the proposition is not *quite* true—it is only approximately true. But everything in geometry, and subsequently in physics, has been derived from it by successive generalizations. The latest of these generalizations is the general theory of relativity.

The theorem of Pythagoras was itself, in all probability, a generalization of an Egyptian rule of thumb. In Egypt, it had been known for ages that a triangle whose sides are 3, 4, and 5 units of length is a right-angled triangle; the Egyptians used this knowledge practically in measuring their fields. Now if the sides of a triangle are 3, 4 and 5 inches, the squares on these sides will contain respectively 9, 16, and 25 square inches; and 9 and 16 added together make 25. Three times three is written '3^2'; four times four, '4^2'; five times five, '5^2.' So that we have

$$3^2 + 4^2 = 5^2$$

It is supposed that Pythagoras noticed this fact, after he had learned from the Egyptians that a triangle whose sides are 3, 4 and 5 has a right angle. He found that this could be generalized, and so arrived at his famous theorem: In a right-angled triangle, the square on the side opposite the right angle is equal to the sum of the squares on the other two sides.

Similarly in three dimensions: if you take a right-angled solid block, the square on the diagonal (the dotted line in the figure) is equal to the sum of the squares on the three sides.

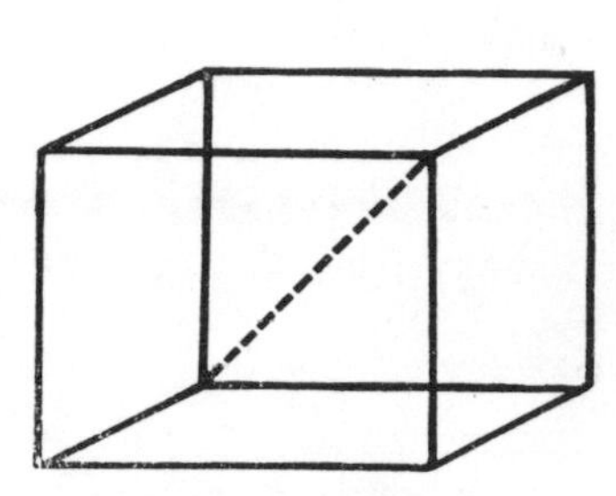

This is as far as the ancients got in this matter.

The next step of importance is due to Descartes, who made the theorem of Pythagoras the basis of his method of analytical geometry. Suppose you wish to map out systematically all the places on a plain—we will suppose it small enough to make it possible to ignore the fact that the earth is round. We will suppose that you live in the middle of the plain. One of the simplest ways of describing the position of a place is to say: starting from my house, go first such and such a distance east, then such and such a distance north (or it may be west in the first case, and south in the second). This tells you exactly where the place is. In the rectangular cities of America, it is the natural method to adopt: in New York you will be told to go so many blocks east (or west) and then so many blocks north (or south). The distance you have to go east is called x, and the distance you have to go north is called y. (If you have to go west, x is negative; if you have to go south, y is negative.) Let O be your starting-point (the 'origin'); let OM be the distance you go east, and MP the distance you go north. How far are you from home in a direct line when you reach P? The theorem of Pythagoras gives the answer. The square on OP is the sum of the squares on OM and MP. If OM is four miles, and MP

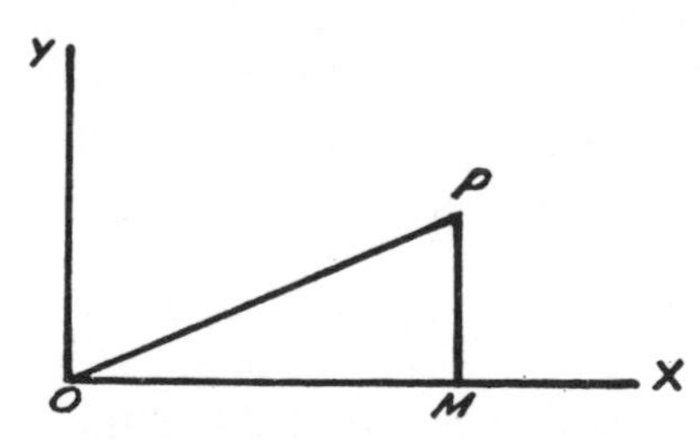

is three miles, OP is five miles. If OM is twelve miles and MP is five miles, OP is thirteen miles, because $12^2 + 5^2 = 13^2$. So that if you adopt Descartes' method of mapping, the theorem of Pythagoras is essential in giving you the distance from place to place. In three dimensions the thing is exactly analogous. Suppose that, instead of wanting merely to fix positions on the plain, you want to fix stations for captive balloons above it, you will then have to add a third quantity, the height at which the balloon is to be. If you call the height z, and if r is the direct distance from O to the balloon, you will have

$$r^2 = x^2 + y^2 + z^2,$$

and from this you can calculate r when you know x, y, and z. For example, if you can get to the balloon by going 12 miles east, 4 miles north, and then 3 miles up, your distance from the balloon in a straight line is thirteen miles, because $12 \times 12 = 144$, $4 \times 4 = 16$, $3 \times 3 = 9$, $144 + 16 + 9 = 169 = 13 \times 13$.

But now suppose that, instead of taking a small piece of the earth's surface which can be regarded as flat, you consider making a map of the world. An accurate map of the world on flat paper is impossible. A globe can be accurate, in the sense that everything is produced to scale, but a flat map cannot be. I am not talking of practical difficulties, I am talking of a theoretical impossibility. For example: the northern halves of the meridian of Greenwich and the 90th meridian of west longitude, together with the piece of the equator between them, make a triangle whose sides are all equal and whose angles are all right angles. On a flat surface, a triangle of that sort would be impossible. On the other hand, it is possible to make a square on a flat surface, but on a sphere it is impossible. Suppose you try on the earth: walk 100 miles west, then 100 miles north, then 100 miles east, then 100 miles south. You might think this would make a square, but it wouldn't, because you would not at the end have come back to your starting-point. If you have time, you may convince yourself of this by experiment. If not, you can easily see that it must be so. When you are nearer the pole, 100 miles takes you through more longitude than when you are nearer the equator, so that in doing your

100 miles east (if you are in the northern hemisphere) you get to a point further east than that from which you started. As you walk due south after this, you remain further east than your starting-point, and end up at a different place from that in which you began. Suppose, to take another illustration, that you start on the equator 4,000 miles east of the Greenwich meridian; you travel till you reach the meridian, then you travel northwards along it for 4,000 miles, through Greenwich and up to the neighbourhood of the Shetland Islands; then you travel eastwards for 4,000 miles, and then 4,000 miles south. This will take you to the equator to a point about 4,000 miles further east than the point from which you started.

In a sense, what we have just been saying is not quite fair, because, except on the equator, travelling due east is not the shortest route from a place to another place due east of it. A ship travelling (say) from New York to Lisbon, which is nearly due east, will start by going a certain distance northward. It will sail on a 'great circle,' that is to say, a circle whose centre is the centre of the earth. This is the nearest approach to a straight line that can be drawn on the surface of the earth. Meridians of longitude are great circles, and so is the equator, but the other parallels of latitude are not. We ought, therefore, to have supposed that, when you reach the Shetland Islands, you travel 4,000 miles, not due east, but along a great circle which lands you at a point due east of the Shetland Islands. This, however, only reinforces our conclusion: you will end at a point even further east of your starting-point than before.

What are the differences between the geometry on a sphere and the geometry on a plane? If you make a triangle on the earth, whose sides are great circles, you will not find that the angles of the triangle add up to two right angles: they will add up to rather more. The amount by which they exceed two right angles is proportional to the size of the triangle. On a small triangle such as you could make with strings on your lawn, or even on a triangle formed by three ships which can just see each other, the angles will add up to so little more than two right angles that you will not be able to detect the difference. But if you take the triangle made by the equator, the Greenwich meridian, and the 90th meridian, the angles add up to *three* right angles. And you can get triangles in which the angles add up to anything up to six right angles. All this you could discover by measurements on the surface of the earth, without having to take account of anything in the rest of space.

The theorem of Pythagoras also will fail for distances on a sphere. From the point of view of a traveller bound to the earth, the distance between two places is their great-circle distance, that is to say, the shortest journey that a man can make without leaving the surface of the earth. Now suppose you take three bits of great circles which make a triangle, and suppose one of them is at right angles to another—to be definite, let one be the equator and one a bit of the meridian of Greenwich going northward from the equator. Suppose you go 3,000 miles along the equator and then 4,000 miles due north; how far will you be from your starting-point, estimating the distance along a great circle? If you were on a plane, your distance would be 5,000 miles, as we saw before. In fact, however, your great-circle distance will be considerably less than this. In a right-angled triangle on a sphere, the square on the side opposite the right angle is less than the sum of the squares on the other two sides.

These differences between the geometry on a sphere and the geometry on a plane are intrinsic differences; that is to say, they enable you to find out whether the surface on which you live is like a plane or like a sphere, without requiring that you should take account of anything outside the surface. Such considerations led to the next step of importance in our subject, which was taken by Gauss, who flourished a hundred and fifty years ago. He studied the theory of surfaces, and showed how to develop it by means of measurements on the surfaces themselves, without going outside them. In order to fix the position of a point in space, we need three measurements; but in order to fix the position of a point on a surface we need only two—for example, a point on the earth's surface is fixed when we know its latitude and longitude.

Now Gauss found that, whatever system of measurement you adopt, and whatever the nature of the surface, there is always a way of calculating the distance between two not very distant points of the surface, when you know the quantities which fix their positions. The formula for the distance is a generalization of the formula of Pythagoras; it tells you the square of the distance in terms of the squares of the differences between the measure-quantities which fix the points, and also the product of these two quantities. When you know this formula, you can discover all the intrinsic properties of the surface, that is to say, all those which do not depend upon its relations to points outside the surface. You can discover, for example, whether the angles of a triangle add up to two right angles, or more, or less, or more in some cases and less in others.

But when we speak of a 'triangle,' we must explain what we mean, because on most surfaces there are no straight lines. On a sphere, we shall replace straight lines by great circles, which are the nearest possible approach to straight lines. In general, we shall take, instead of straight lines, the lines that give the shortest route on the surface from place to place. Such lines are called 'geodesics.' On the earth, the geodesics are great circles. In general, they are the shortest way of travelling from point to point if you are unable to leave the surface. They take the place of straight lines in the intrinsic geometry of a surface. When we inquire whether the angles of a triangle add up to two right angles or not, we mean to speak of a triangle whose sides are geodesics. And when we speak of the distance between two points, we mean the distance along a geodesic.

The next step in our generalizing process is rather difficult: it is the transition to non-Euclidean geometry. We live in a world in which space has three dimensions, and our empirical knowledge of space is based upon measurement of small distances and of angles. (When I speak of small distances I mean distances that are small compared to those in astronomy; all distances on the earth are small in this sense.) It was formerly thought that we could be sure *a priori* that space is Euclidean—for instance, that the angles of a triangle add up to two right angles. But it came to be recognized that we could not prove this by reasoning; if it was to be known, it must be known as the result of measurements. Before Einstein, it was thought that measurements confirm Euclidean geometry within the limits of exactitude attainable; now this is no longer thought. It is still true that we can, by what may be called a natural artifice, cause Euclidean geometry to *seem* true throughout a small region, such as the earth; but in explaining gravitation Einstein is led to the view that over large regions where there is matter we cannot regard space as Euclidean. The reasons for this will concern us later. What concerns us now is the way in which non-Euclidean geometry results from a generalization of the work of Gauss.

There is no reason why we should not have the same circumstances in three-dimensional space as we have, for example, on the surface of a sphere. It might happen that the angles of a triangle would always add up to more than two right angles, and that the excess would be proportional to the size of the triangle. It might happen that the distance between two points would be given by a formula analogous to what we have on the surface of a sphere, but involving three quantities instead of two. Whether this does happen or not, can only be discovered by actual measurements. There are an infinite number of such possibilities.

This line of argument was developed by Riemann, in his dissertation 'On the hypotheses which underlie geometry' (1854), which applied Gauss's work on surfaces to different kinds of three-dimensional spaces. He showed that all the essential characteristics of a kind of space could be deduced from the formula for small distances. He assumed that, from the small distances in three given directions which would together carry you from one point to another not far from it, the distances between the two points could be calculated. For instance, if you know that you can get from one point to another by first moving a certain distance east, then a certain distance north, and finally a certain distance straight up in the air, you are to be able to calculate the distance from the one point to the other. And the rule for the calculation is to be an extension of the theorem of Pythagoras, in the sense that you arrive at the square of the required distance by adding together multiples of the squares of the component distances, together possibly with multiples of their products. From certain characteristics in the formula, you can tell what sort of space you have to deal with. These characteristics do not depend upon the particular method you have adopted for determining the positions of points.

In order to arrive at what we want for the theory of relativity, we now have one more generalization to make: we have to substitute the 'interval' between events for the distance between points. This takes us to space-time. We have already seen that, in the special theory of relativity, the square of the interval is found by subtracting the square of the distance between events from the square of the distance that light would travel in the time between them. In the general theory, we do not assume this special form of interval. We assume to begin with a general form, like that which Riemann used for distances. Moreover, like Riemann, Einstein only assumes his formula for *neighbouring* events, that is to say, events which have only a small interval between them. What goes beyond these initial assumptions depends upon observation of the actual motion of bodies, in ways which we shall explain in later chapters.

We may now sum up and re-state the process we have been describing. In three dimensions, the position of a point relatively to a fixed point (the 'origin') can be determined by assigning three quantities ('co-ordinates'). For example, the position of a balloon relatively to your house is fixed if you know that you will reach it by going first a given distance due east, then another given distance due north, then a third given distance straight up. When, as in this case, the three co-ordinates are three distances all at right angles to each other, which, taken successively, transport you from the origin to the point in question, the square of the direct distance to the point in question is got by adding up the squares of the three co-ordinates. In all cases, whether in Euclidean or in non-Euclidean spaces, it is got by adding multiples of the squares and products of the co-ordinates according to an assignable rule. The co-ordinates may be any quantities which fix the position of a point, provided that neighbouring points must have neighbouring quantities for their co-ordinates. In the general theory of relativity, we add a fourth co-ordinate to give the time, and our formula gives 'interval' instead of spatial distance; moreover we assume the accuracy of our formula for small distances only.

We are now at last in a position to tackle Einstein's theory of gravitation.

CHAPTER VIII

Einstein's Law of Gravitation

BEFORE tackling Einstein's new law, it is as well to convince ourselves, on logical grounds, that Newton's law of gravitation cannot be quite right.

Newton said that between any two particles of matter there is a force which is proportional to the product of their masses and inversely proportional to the square of their distance. That is to say, ignoring for the present the question of mass, if there is a certain attraction when the particles are a mile apart, there will be a quarter as much attraction when they are two miles apart, a ninth as much when they are three miles apart, and so on: the attraction diminishes much faster than the distance increases. Now, of course, Newton, when he spoke of the distance, meant the distance at a given time: he thought there could be no ambiguity about time. But we have seen that this was a mistake. What one observer judges to be the same moment on the earth and the sun, another will judge to be two different moments. 'Distance at a given moment' is therefore a subjective conception, which can hardly enter into a cosmic law. Of course, we could make our law unambiguous by saying that we are going to estimate times as they are estimated by Greenwich Observatory. But we can hardly believe that the accidental circumstances of the earth deserve to be taken so seriously. And the estimate of distance, also, will vary for different observers. We cannot therefore allow that Newton's form of the law of gravitation can be quite correct, since it will give different results according to which of many equally legitimate conventions we adopt. This is as absurd as it would be if the question whether one man had murdered another were to depend upon whether they were described by their Christian names or their surnames. It is obvious that physical laws must be the same whether distances are measured in miles or in kilometres, and we are concerned with what is essentially only an extension of the same principle.

Our measurements are conventional to an even greater extent than is admitted by the special theory of relativity. Moreover, every measurement is a physical process carried out with physical material; the result is certainly an experimental datum, but may not be susceptible of the simple interpretation which we ordinarily assign to it. We are, therefore, not going to assume to begin with that we know how to measure anything. We assume that there is a certain physical quantity called 'interval,' which is a relation between two events that are not widely separated; but we do not assume in advance that we know how to measure it, beyond taking it for granted that it is given by some generalization of the theorem of Pythagoras such as we spoke of in the preceding chapter.

We do assume, however, that events have an *order*, and that this order is four-dimensional. We assume, that is to say, that we know what we mean by saying that a certain event is nearer to another than a third, so that before making accurate measurements we can speak of the 'neighbourhood' of an event; and we assume that, in order to assign the position of an event in space-time, four quantities (co-ordinates) are necessary—e.g., in our former case of an explosion on an airship, latitude, longitude, altitude and time. But we assume nothing about the way in which these co-ordinates are assigned, except that neighbouring co-ordinates are assigned to neighbouring events.

The way in which these numbers, called co-ordinates, are to be assigned is neither wholly arbitrary nor a result of careful measurement—it lies in an intermediate region. While you are making any continuous journey, your co-ordinates must never alter by sudden jumps. In America one finds that the houses between (say) 14th Street and 15th Street are likely to have numbers between 1400 and 1500, while those between 15th Street and 16th Street have numbers between 1500 and 1600, even if the 1400's were not used up. This would not do for our purposes, because there is a sudden jump when we pass from one block to the next. Or again we might assign the time co-ordinate in the following way: take the time that elapses between two successive births of people called Smith; an event occurring between the births of the 3000th and the 3001st Smith known to history shall have a co-ordinate lying between 3000 and 3001; the fractional part of its co-ordinate shall be the fraction of a year that has elapsed since the birth of the 3000th Smith. (Obviously there could never be as much as a year between two successive additions to the Smith family.) This way of assigning the time co-ordinate is perfectly definite, but it is not admissible for our purposes, because there will be sudden jumps between events just before the birth of a Smith and events just after, so that in a continuous journey your time co-ordinate will not change continuously. It is assumed that, independently of measurement, we know what a continuous journey is. And when your position in space-time changes continuously, each of your four co-ordinates must change continuously. One, two or three of them may not change at all; but whatever change does occur must be smooth, without sudden jumps. This explains what is *not* allowable in assigning co-ordinates.

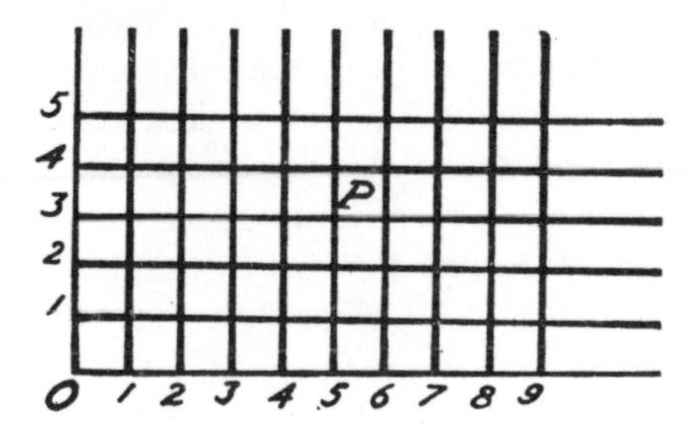

To explain all the changes that are legitimate in your co-ordinates, suppose you take a large piece of soft india-rubber. While it is in an unstretched condition, measure little squares on it, each one-tenth of an inch each way. Put in little tiny pins at the corners of the squares. We can take as two of the co-ordinates of one of these pins the number of pins passed in going to the right from a given pin until we come just below the pin in question, and then the number of pins we pass on the way up to this pin. In the figure, let O be the pin we start from and P the pin to which we are going to assign co-ordinates.

P is in the 5th column and the 3rd row, so its co-ordinates in the plane of the india-rubber are to be 5 and 3.

Now take the india-rubber and stretch it and twist it as

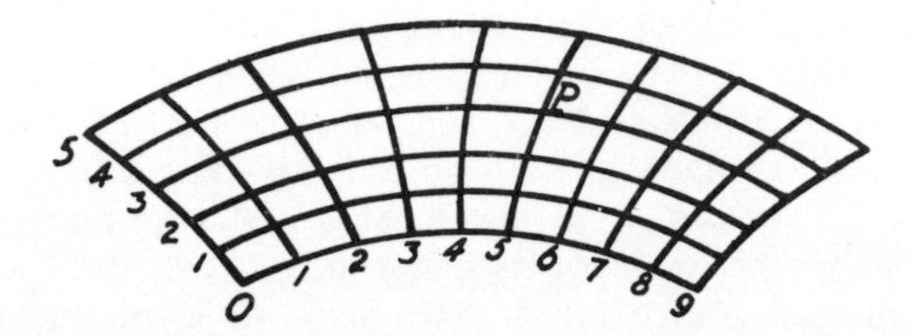

much as you like. Let the pins now be in the shape they have in the second figure. The divisions now no longer represent distances according to our usual notions, but they will still do just as well as co-ordinates. We may still take P as having the co-ordinates 5 and 3 in the plane of the india-rubber; and we may still regard the india-rubber as being in a plane, even if we have twisted it out of what we should ordinarily call a plane. Such continuous distortions do not matter.

To take another illustration: instead of using a steel measuring-rod to fix our co-ordinates, let us use a live eel, which is wriggling all the time. The distance from the tail to the head of the eel is to count as 1 from the point of view of co-ordinates, whatever shape the creature may be assuming at the moment. The eel is continuous, and its wriggles are continuous, so it may be taken as our unit of distance in assigning co-ordinates. Beyond the requirement of continuity, the method of assigning co-ordinates is purely conventional, and therefore a live eel is just as good as a steel rod.

We are apt to think that, for really careful measurements, it is better to use a steel rod than a live eel. This is a mistake; not because the eel tells us what the steel rod was thought to tell, but because the steel rod really tells no more than the eel obviously does. The point is, not that eels are really rigid, but that steel rods really wriggle. To an observer in just one possible state of motion the eel would appear rigid, while the steel rod would seem to wriggle just as the eel does to us. For everybody moving differently both from this observer and ourselves, both the eel and the rod would seem to wriggle. And there is no saying that one observer is right and another wrong. In such matters what is seen does not belong solely to the physical process observed, but also to the standpoint of the observer. Measurements of distances and times do not directly reveal properties of the things measured, but relations of the things to the measurer. What observation can tell us about the physical world is therefore more abstract than we have hitherto believed.

It is important to realize that geometry, as taught in schools since Greek times, ceases to exist as a separate science, and becomes merged into physics. The two fundamental notions in elementary geometry were the straight line and the circle. What appears to you as a straight road, whose parts all exist now, may appear to another observer to be like the flight of a rocket, some kind of curve whose parts come into existence successively. The circle depends upon measurement of distances, since it consists of all the points at a given distance from its centre. And measurement of distances, as we have seen, is a subjective affair, depending upon the way in which the observer is moving. The failure of the circle to have objective validity was demonstrated by the Michelson–Morley experiment, and is thus, in a sense, the starting-point of the whole theory of relativity. Rigid bodies, which we need for measurement, are only rigid for certain observers; for others they will be constantly changing all their dimensions. It is only our obstinately earth-bound imagination that makes us suppose a geometry separate from physics to be possible.

That is why we do not trouble to give physical significance to our co-ordinates from the start. Formerly, the co-ordinates used in physics were supposed to be carefully measured distances; now we realize that this care at the start is thrown away. It is at a later stage that care is required. Our co-ordinates now are hardly more than a systematic way of cataloguing events. But mathematics provides, in the method of tensors, such an immensely powerful technique that we can use co-ordinates assigned in this apparently careless way just as effectively as if we had applied the whole apparatus of minutely accurate measurement in arriving at them. The advantage of being haphazard at the start is that we avoid making surreptitious physical assumptions, which we can hardly help making if we suppose that our co-ordinates have initially some particular physical significance.

We need not try to proceed in ignorance of all observed physical phenomena. We know certain things. We know that the old Newtonian physics is very nearly accurate when our co-ordinates have been chosen in a certain way. We know that the special theory of relativity is still more nearly accurate for suitable co-ordinates. From such facts we can infer certain things about our new co-ordinates, which, in a logical deduction, appear as postulates of the new theory.

As such postulates we take:

1. That the interval between neighbouring events takes a general form, like that used by Riemann for distances.

2. That every body travels on a geodesic in space-time, except in so far as non-gravitational forces act upon it.

3. That a light-ray travels on a geodesic which is such that the interval between any two parts of it is zero.

Each of these postulates requires some explanation.

Our first postulate requires that, if two events are close together (but not necessarily otherwise), there is an interval between them which can be calculated from the differences between their co-ordinates by some such formula as we considered in the preceding chapter. That is to say, we take the squares and products of the differences of co-ordinates, we multiply them by suitable amounts (which in general will vary from place to place), and we add the results together. The sum obtained is the square of the interval. We do not assume in advance that we know the amounts by which the squares and products must be multiplied; this is going to be discovered by observing physical phenomena. But we do know, because Riemann's mathematics shows it to be so, that within any small region of space-time we can choose the co-ordinates so that the interval has almost exactly the special form which we found in the special theory of relativity. It is not necessary for the application of the special theory to a limited region that there should be no gravitation in the region; it is enough if the intensity of gravitation is practically the same throughout the region. This enables us to apply the special theory within any small region. How small it will have to be, depends upon the neighbourhood. On the surface of the earth, it would have to be small enough for the curvature of the earth to be negligible. In the spaces between the planets, it need only be small enough for the attraction of the sun and the planets to be sensibly constant throughout the region. In the spaces between the stars it might be enormous—say half the distance from one star to the next—without introducing measurable inaccuracies.

Thus, at a great distance from gravitating matter, we can so choose our co-ordinates as to obtain very nearly a Euclidean space; this is really only another way of saying that the special theory of relativity applies. In the neighbourhood of matter, although we can still make our space very nearly Euclidean in a very small region, we cannot do so throughout any region within which gravitation varies sensibly—at least, if we do, we shall have to abandon the view expressed in the second postulate, that bodies moving under gravitational forces only move on geodesics.

We saw that a geodesic on a surface is the shortest line that can be drawn on the surface from one point to another; for example, on the earth the geodesics are great circles. When we come to space-time, the mathematics is the same, but the verbal explanations have to be rather different. In the general theory of relativity, it is only neighbouring events that have a definite interval, independently of the route by which we travel from one to the other. The interval between distant events depends upon the route pursued, and has to be calculated by dividing the route into a number of little bits and adding up the intervals for the various little bits. If the interval is space-like, a body cannot travel from one event to the other; therefore when we are considering the way bodies move, we are confined to time-like intervals. The interval between neighbouring events when it is time-like, will appear as the time between them for an observer who travels from the one event to the other. And so the whole interval between two events will be judged by a person who travels from one to the other to be what his clocks show to be the time that he has taken on the journey. For some routes this time will be longer, for others shorter; the more slowly the man travels, the longer he will think he has been on the journey. This must not be taken as a platitude. I am not saying that if you travel from London to Edinburgh you will take longer if you travel more slowly. I am saying something much more odd. I am saying that if you leave London at 10 a.m. and arrive in Edinburgh at 6.30 p.m., Greenwich time, the more slowly you travel the longer you will take—if the time is judged by your watch. This is a very different statement. From the point of view of a person on the earth, your journey takes eight hours and a half. But if you had been a ray of light travelling round the solar system, starting from London at 10 a.m., reflected from Jupiter to Saturn, and so on, until at last you were reflected back to Edinburgh and arrived there at 6.30 p.m., you would judge that the journey had taken you exactly no time. And if you had gone by any circuitous route, which enabled you to arrive in time by travelling fast, the longer your route the less time you would judge that you had taken; the diminution of time would be continual as your speed approached that of light. Now I say that when a body travels, if it is left to itself, it chooses the route which makes the time between two stages of the journey as long as possible; if it had travelled from one event to another by any other route, the time, as measured by its own clocks, would have been shorter. This is a way of saying that bodies left to themselves do their journeys as slowly as they can; it is a sort of law of cosmic laziness. Its mathematical expression is that they travel in geodesics, in which the total interval between any two events on the journey is *greater* than by any alternative route. (The fact that it is greater, not less, is due to the fact that the sort of interval we are considering is more analogous to time than to distance.) For example, if a person could leave the earth and travel about for a time and then return, the time between his departure and return would be less by his clocks than by those on the earth: the earth, in its journey round the sun, chooses the route which makes the time of any bit of its course by its clocks longer than the time as judged by clocks which move by a different route. This is what is meant by saying that bodies left to themselves move in geodesics in space-time.

It is important to remember that space-time is not supposed to be Euclidean. As far as the geodesics are concerned, this has the effect that space-time is like a hilly countryside. In the neighbourhood of a piece of matter, there is, as it were, a hill in space-time; this hill grows steeper and steeper as it gets nearer the top, like the neck of a champagne bottle. It ends in a sheer precipice. Now by the law of cosmic laziness which we mentioned earlier, a body coming into the neighbourhood of the hill will not attempt to go straight over the top, but will go round. This is the essence of Einstein's view of gravitation. What a body does, it does because of the nature of space-time in its own neighbourhood, not because of some mysterious force emanating from a distant body.

An analogy will serve to make the point clear. Suppose that on a dark night a number of men with lanterns were walking in various directions across a huge plain, and suppose that in one part of the plain there was a hill with a flaring beacon on the top. Our hill is to be such as we have described, growing steeper, as it goes up and ending in a precipice. I shall suppose that there are villages dotted about the plain, and the men with lanterns are walking to and from these various villages. Paths have been made showing the easiest way from any one village to any other. These paths will all be more or less curved, to avoid going too far up the hill; they will be more sharply curved when they pass near the top of the hill than when they keep some way off from it. Now suppose that you are observing all this, as best you can, from a place high up in a balloon, so that you cannot see the ground, but only the lanterns and the beacon. You will not know that there is a hill, or that the beacon is at the top of it. You will see that people turn out of the straight course when they approach the beacon, and that the nearer they come the more they turn aside. You will naturally attribute this to an effect of the beacon; you may think that it is very hot and people are afraid of getting burnt. But if you wait for daylight you will see the hill, and you will find that the beacon merely marks the top of the hill and does not influence the people with lanterns in any way.

Now in this analogy the beacon corresponds to the sun, the people with lanterns correspond to the planets and comets, the paths correspond to their orbits, and the coming of daylight corresponds to the coming of Einstein. Einstein says that the sun is at the top of a hill, only the hill is in space-time, not in space. (I advise the reader not to try to picture this, because it is impossible.) Each body, at each moment, adopts the easiest course open to it, but owing to the hill the easiest course is not a straight line. Each little bit of matter is at the top of its own little hill, like the cock on his own dung-heap. What we call a big bit of matter is a bit which is the top of a big hill. The hill is what we know about; the bit of matter at the top is assumed for convenience. Perhaps there is really no need to assume it, and we could do with the hill alone, for we can never get to the top of anyone else's hill, any more than the pugnacious cock can fight the peculiarly irritating bird that he sees in the looking-glass.

I have given only a qualitative description of Einstein's law of gravitation; to give its exact quantitative formulation is impossible without more mathematics than I am permitting myself. The most interesting point about it is that it makes the law no longer the result of action at a distance; the sun exerts no force on the planets whatever. Just as geometry has become physics, so, in a sense, physics has become geometry. The law of gravitation has become the geometrical law that every body pursues the easiest course from place to place, but this course is affected by the hills and valleys that are encountered on the road.

We have been assuming that the body considered is acted upon only by gravitational forces. We are concerned at present with the law of gravitation, not with the effects of electro-magnetic forces or the forces between sub-atomic particles. There have been many attempts to bring all these forces into the framework of general relativity, by Einstein himself, and by Weyl, Kaluza and Klein, to mention only a few of the others, but none of these attempts has been entirely satisfactory. For the present, we may ignore this work, because the planets are not subject, as wholes, to appreciable electromagnetic or sub-atomic forces; it is only gravitation that has to be considered in accounting for their motions, with which we are concerned in this chapter.

Our third postulate, that a light-ray travels so that the interval between two parts of it is zero, has the advantage that it does not have to be stated only for *small* distances. If each little bit of interval is zero, the sum of them all is zero, and so even distant parts of the same light-ray have a zero interval. The course of a light-ray is also a geodesic according to this postulate. Thus we now have two empirical ways of discovering what are the geodesics in space-time, namely light-rays and bodies moving freely. Among freely-moving bodies are included all which are not subject as wholes, to appreciable electromagnetic or sub-atomic forces, that is to say, the sun, stars, planets and satellites, and also falling bodies on the earth, at least when they are falling in a vacuum. When you are standing on the earth, you are subject to electromagnetic forces: the electrons and protons in the neighbourhood of your feet exert a repulsion on your feet which is just enough to overcome the earth's gravitation. This is what prevents you from falling through the earth, which, solid as it looks, is mostly empty space.

CHAPTER IX

Proofs of Einstein's Law of Gravitation

THE reasons for accepting Einstein's law of gravitation rather than Newton's are partly empirical, partly logical. We will begin with the former.

Einstein's law of gravitation gives very nearly the same results as Newton's, when applied to the calculation of the orbits of the planets and their satellites. If it did not, it could not be true, since the consequences deduced from Newton's law have been found to be almost exactly verified by observation. When, in 1915, Einstein first published his new law, there was only one empirical fact to which he could point to show that his theory was better than Newton's. This was what is called the motion of the perihelion of Mercury.

The planet Mercury, like the other planets, moves round the sun in an ellipse, with the sun in one of the foci. At some points of its orbit it is nearer to the sun than at other points. The point where it is nearest to the sun is called its 'perihelion.' Now it was found by observation that, from one occasion when Mercury is nearest to the sun until the next, Mercury does not go exactly once round the sun, but a little bit more. The discrepancy is very small; it amounts to an angle of forty-two seconds in a century. Since Mercury goes round the sun rather more than four hundred times a century, it must move about one-tenth of a second of angle more than a complete revolution to get from one perihelion to the next. This very minute discrepancy from Newtonian theory had puzzled astronomers. There was a calculated effect due to perturbations caused by the other planets, but this small discrepancy was the residue after allowing for these perturbations. Einstein's theory accounted exactly for this residue. There is a similar effect in the case of the other planets, but it is much smaller and more difficult to observe. Since Einstein published his new law, the effect has also been observed for the earth, and with a fair degree of certainty for Mars. This perihelion effect was, at first, Einstein's only empirical advantage over Newton.

His second success was more sensational. According to orthodox opinion, light in a vacuum ought always to travel in straight lines. Not being composed of material particles, it ought to be unaffected by gravitation. However, it was possible without any serious breach with old ideas, to admit that, in passing near the sun, light might be deflected out of the straight path as much as if it were composed of material particles. Einstein, however, maintained, as a deduction from his law of gravitation, that light would be deflected twice as much as this. That is to say, if the light of a star passed very near the sun, Einstein maintained that the ray from the star would be turned through an angle of just under one second and three-quarters. His opponents were willing to concede half of this amount. Unfortunately, stars which are almost in line with the sun can be seen only during a total eclipse, and even then there may be no sufficiently bright stars near to the sun. Eddington points out that, from this point of view, the best day of the year is May 29 because then there are a number of bright stars close to the sun. It happened by incredible good fortune that there was a total eclipse of the sun on May 29, 1919. Two British expeditions photographed the stars near the sun during the eclipse, and the results appeared to confirm Einstein's prediction. This caused great excitement at the time. Supporting evidence has been obtained from observation of many subsequent eclipses, and Einstein's prediction is therefore generally accepted. However there are many possible sources of error in such observations, and some astronomers still doubt that the results are quite conclusive.

The third experimental test is on the whole favourable to Einstein, but the quantities concerned are so small that it is only just possible to measure them, and the result is therefore not decisive. Before explaining the effect in question, a few preliminary explanations are necessary. The spectrum of an element consists of certain lines of various shades of light, emitted by the element when it glows, which may be separated by a prism. They are the same (to a close approximation) whether the element is on the earth, the sun, or a star. Each line is of some definite shade of colour, with some definite wave-length. Longer wave-lengths are towards the red end of the spectrum, shorter ones towards the violet end. When the source of light is moving towards you, the apparent wave-lengths grow shorter, just as waves at sea come quicker when you are travelling against the wind. When the source of light is moving away from you, the apparent wave-lengths grow longer, for the same reason. This enables us to know whether the stars are moving towards us or away from us. If they are moving towards us, all the lines in the spectrum of an element are moved a little towards violet; if away from us, towards red. You may notice the analogous effect in sound any day. If you are in a station and an express comes through whistling, the note of the whistle seems much more shrill while the train is approaching you than when it has passed. Probably many people think the note has 'really' changed, but in fact the change in what you hear is only due to the fact that the train was first approaching and then receding. To people in the train, there was no change of note. This is *not* the effect with which Einstein is concerned. The distance of the sun from the earth does not change much; for our present purposes, we may regard it as constant. Einstein deduces from his law of gravitation that any periodic process which takes place in an atom in the sun (whose gravitation is very intense) must, as measured by our clocks, take place at a slightly slower rate than it would in a similar atom on the earth. The 'interval' involved will be the same in the sun and on the earth, but the same interval in different regions does not correspond to exactly the same time; this is due to the 'hilly' character of space-time which constitutes gravitation. Consequently any given line in the spectrum ought, when the light comes from the sun, to seem to us a little nearer the red end of the spectrum than if the light came from a source on the earth. Einstein's theory predicts that a similar effect should be observed in the gravitational field of every star, and indeed of any massive body, but the effect is so small and the difficulty of measuring it so great that after fifty years of observation it is still uncertain whether the sun or any star exhibits the effect at all. However, recent advances in nuclear physics have made it possible to observe the effect produced by the earth itself, although the earth's effect is much smaller than the sun's. The new observation depends on the fact that under suitable experimental conditions, certain radioactive nuclei emit electromagnetic radiation whose wavelength can be determined with immense precision, and that very small changes in wavelength may also be detected. The change in wavelength due to gravitation has now been observed between two stations, one at the top and the other at the bottom of a tower only 74 feet high.

There are other differences between the consequences of Einstein's law and of Newton's, but hitherto there has been no other decisive observation, at least within the limits of the solar system. But the above experimental tests are quite sufficient to convince astronomers that, where Newton and Einstein differ as to the motions of the heavenly bodies, it is Einstein's law that gives the right results. Even if the empirical grounds in favour of Einstein stood alone, they would be conclusive. Whether his law represents the exact truth or not, it is certainly more nearly exact than Newton's, though the inaccuracies in Newton's were all exceedingly minute.

But the considerations which originally led Einstein to his law were not of this detailed kind. Even the consequence about the perihelion of Mercury, which could be verified at once from previous observations, could only be deduced after the theory was complete, and could not form any part of the original grounds for inventing such a theory. These grounds were of a more abstract logical character. I do not mean that they were not based upon observed facts, and I do not mean that they were *a priori* fantasies such as philosophers indulged in formerly. What I mean is that they were derived from certain general characteristics of physical experience, which showed that Newton *must* be wrong and that something like Einstein's law *must* be substituted.

The arguments in favour of the relativity of motion are, as we saw in earlier chapters, quite conclusive. In daily life, when we say that something moves, we mean that it moves relatively to the earth. In dealing with the motions of the plants, we consider them as moving relatively to the sun, or to the centre of mass of the solar system. When we say that the solar system itself is moving, we mean that it is moving relatively to the stars. There is no physical occurrence which can be called 'absolute motion.' Consequently the laws of physics must be concerned with relative motions, since these are the only kind that occur.

We now take the relativity of motion in conjunction with the experimental fact that the velocity of light is the same relatively to one body as relatively to another, however the two may be moving. This leads us to the relativity of distances and times. This in turn shows that there is no objective physical fact which can be called 'the distance between two bodies at a given time,' since the time and the distance will both depend on the observer. Therefore Newton's law of gravitation is logically untenable, since it makes use of 'distance at a given time.'

This shows that we cannot rest content with Newton, but it does not show what we are to put in his place. Here several considerations enter in. We have in the first place what is called 'the equality of gravitational and inertial mass.' What this means is as follows: When you apply a given force[1] to a heavy body, you do not give it as much acceleration as you would to a light body. What is called the 'inertial' mass of a body is measured by the amount of force required to produce a given acceleration. At a given point of the earth's surface, the 'mass' is proportional to the 'weight'. What is measured by scales is rather the mass than the weight: the weight is defined as the force with which the earth attracts the body. Now this force is greater at the poles than at the equator, because at the equator the rotation of the earth produces a 'centrifugal force' which partially counteracts gravitation. The force of the earth's attraction is

[1] Although 'force' is no longer to be regarded as one of the fundamental concepts of dynamics, but only as a convenient way of speaking, it can still be employed like 'sunrise' and 'sunset', provided we realize what we mean. Often it would require very roundabout expressions to avoid the term 'force.'

also greater on the surface of the earth than it is at a great height or at the bottom of a very deep mine. None of these variations are shown by scales, because they affect the weights used just as much as the body weighed: but they are shown if we use a spring balance. The mass does not vary in the course of these changes of weight.

The 'gravitational' mass is differently defined. It is capable of two meanings. We may mean (1) the way a body responds in a situation where gravitation has a known intensity, for example, on the surface of the earth, or on the surface of the sun; or (2), the intensity of the gravitational force produced by the body, as, for example, the sun produces stronger gravitational forces than the earth does. Newton says that the force of gravitation between two bodies is proportional to the product of their masses. Now let us consider the attraction of different bodies to one and the same body, say the sun. Then different bodies are attracted by forces which are proportional to their masses, and which, therefore, produce exactly the same acceleration in all of them. Thus if we mean 'gravitational mass' in sense (1), that is to say, the way a body responds to gravitation, we find that 'the equality of inertial and gravitational mass,' which sounds formidable, reduces to this: that in a given gravitational situation, all bodies behave exactly alike. As regards the surface of the earth, this was one of the first discoveries of Galileo. Aristotle thought that heavy bodies fall faster than light ones; Galileo showed that this is not the case, when the resistance of the air is eliminated. In a vacuum, a feather falls as fast as a lump of lead. As regards the planets, it was Newton who established the corresponding facts. At a given distance from the sun, a comet, which has a very small mass, experiences exactly the same acceleration towards the sun as a planet experiences at the same distance. Thus the way in which gravitation affects a body depends only upon where the body is, and in no degree upon the nature of the body. This suggests that the gravitational effect is a characteristic of the locality, which is what Einstein makes it.

As for the gravitational mass in sense (2), i.e. the intensity of the force produced by a body, this is no longer *exactly* proportional to its inertial mass. The question involves some rather complicated mathematics and I shall not go into it.[1]

We have another indication as to what sort of thing the law of gravitation *must* be, if it is to be a characteristic of a neighbourhood, as we have seen reason to suppose that it is. It must be expressed in some law which is unchanged when we adopt a different kind of co-ordinates. We saw that we must not, to begin with, regard our co-ordinates as having any physical significance: they are merely systematic ways of naming different parts of space-time. Being conventional, they cannot enter into physical laws. That means to say that, if we have expressed a law correctly in terms of one set of co-ordinates, it must be expressed by the same formula in terms of another set of co-ordinates. Or, more exactly, it must be possible to find a formula which expresses the law, and which is unchanged however we change the co-ordinates. It is the business of the theory of tensors to deal with such formulae. And the theory of tensors shows that there is one formula which obviously suggests itself as being possibly the law of gravitation. When this possibility is examined, it is found to give the right results; it is here that the empirical confirmations come in. But if Einstein's law had not been found to agree with experience, we could not have gone back to Newton's law. We should have been compelled by logic to seek some law expressed in terms of 'tensors,' and therefore independent of our choice of co-ordinates. It is impossible without mathematics to explain the theory of tensors; the non-mathematician must be content to know that it is the technical method by which we eliminate the conventional element from our measurements and laws, and thus arrive at physical laws which are independent of the observer's point of view. Of this method, Einstein's law of gravitation is the most splendid example.

[1] See Eddington, *The Mathematical Theory of Relativity*, Second Edition, p. 128.

CHAPTER X

Mass, Momentum, Energy, and Action

THE pursuit of quantitative precision is as arduous as it is important. Physical measurements are made with extraordinary exactitude; if they were made less carefully, such minute discrepancies as form the experimental data for the theory of relativity could never be revealed. Mathematical physics, before the coming of relativity, used a set of conceptions which were supposed to be as precise as physical measurements, but it has turned out that they were logically defective, and that this defectiveness showed itself in very small deviations from expectations based upon calculation. In this chapter I want to show how the fundamental ideas of pre-relativity physics are affected, and what modifications they have had to undergo.

We have already had occasion to speak of mass. For purposes of daily life, mass is much the same as weight; the usual measures of weight—ounces, grams, etc.—are really measures of mass. But as soon as we begin to make accurate measurements, we are compelled to distinguish between mass and weight. Two different methods of weighing are in common use, one, that of scales, the other that of the spring balance. When you go a journey and your luggage is weighed, it is not put on scales, but on a spring; the weight depresses the spring a certain amount, and the result is indicated by a needle on a dial. The same principle is used in automatic machines for finding your weight. The spring balance shows weight, but scales show *mass*. So long as you stay in one part of the world, the difference does not matter; but if you test two weighing machines of different kinds in a number of different places, you will find, if they are accurate, that their results do not always agree. Scales will give the same result anywhere, but a spring balance will not. That is to say, if you have a lump of lead weighing 10 lbs. by the scales, it will also weigh 10 lbs. by scales in any other part of the world. But if it weighs 10 lbs. by a spring balance in London, it will weigh more at the North Pole, less at the equator, less high up in an aeroplane, and less at the bottom of a coal-mine, if it is weighed in all those places on the same spring balance. The fact is that the two instruments measure quite different quantities. The scales measure what may be called (apart from refinements which will concern us presently) 'quantity of matter.' There is the same 'quantity of matter' in a pound of feathers as in a pound of lead. Standard 'weights,' which are really standard 'masses,' will measure the amount of mass in any substance put into the opposite scales. But 'weight' is a property due to the earth's gravitation: it is the amount of the force by which the earth attracts a body. This force varies from place to place. In the first place, anywhere outside the earth the attraction varies inversely as the square of the distance from the centre of the earth; it is therefore less at great heights. In the second place, when you go down a coal-mine part of the earth is above you, and attracts matter upwards instead of downwards, so that the net attraction downwards is less than on the surface of the earth. In the third place, owing to the rotation of the earth, there is what is called a 'centrifugal force,' which acts against gravitation. This is greatest at the equator, because there the rotation of the earth involves the fastest motion; at the poles it does not exist, because they are on the axis of rotation. For all these reasons, the force with which a given body is attracted to the earth is measurably different at different places. It is this force that is measured by a spring balance; that is why a spring balance gives different results in different places. In the case of scales, the standard 'weights' are altered just as much as the body to be weighed, so that the result is the same everywhere; but the result is the 'mass,' not the 'weight.' A standard 'weight' has the same mass everywhere, but not the same 'weight'; it is in fact a unit of mass, not of weight. For theoretical purposes, mass, which is almost invariable for a given body, is much more important than weight, which varies according to circumstances. Mass may be regarded, to begin with, as 'quantity of matter'; we shall see that this view is not strictly correct, but it will serve as a starting-point for subsequent refinements.

For theoretical purposes, mass is defined as being determined by the amount of force required to produce a given acceleration: the more massive a body is, the greater will be the force required to alter its velocity by a given amount in a given time. It takes a more powerful engine to make a long train attain a speed of ten miles an hour at the end of the first half-minute, than it does to make a short train do so. Or we may have circumstances where the force is the same for a number of different bodies; in that case, if we can measure the accelerations produced in them, we can tell the ratios of their masses: the greater the mass, the smaller the acceleration. We may take, in illustration of this method, an example which is important in connection with relativity. Radio-active bodies emit beta-particles (electrons) with enormous velocities. We can observe their path by making them travel through water vapour and form a cloud as they go. We can at the same time subject them to known electric and magnetic forces, and observe how much they are bent out of a straight line by these forces. This makes it possible to compare their masses. It is found that the faster they travel, the greater are their masses, as measured by the stationary observer. It is known otherwise that, apart from the effect of motion, all electrons have the same mass.

All this was known before the theory of relativity was invented, but it showed that the traditional conception of mass had not quite the definiteness that had been ascribed to it. Mass used to be regarded as 'quantity of matter,' and supposed to be quite invariable. Now mass was found to be relative to the observer, like length and time, and to be altered by motion in exactly the same proportion. However, this could be remedied. We could take the 'proper mass,' the mass as measured by an observer who shares the motion of the body. This was easily inferred from the measured mass, by taking the same proportion as in the case of lengths and times.

But there is a more curious fact, and that is, that after we have made this correction we still have not obtained a quantity which is at all times exactly the same for the same body.

When a body absorbs energy—for example, by growing hotter—its 'proper mass' increases slightly. The increase is very slight, since it is measured by dividing the increase of energy by the square of the velocity of light. On the other hand, when a body parts with energy it loses mass. The most notable case of this is that four hydrogen atoms can come together to make one helium atom, but a helium atom has rather less than four times the mass of one hydrogen atom. This phenomenon is of the greatest practical importance. It is thought to occur in the interior of stars, providing the energy which we see as starlight and which, in the case of the sun, supports terrestrial life. It can be also made to occur in terrestrial laboratories, with an enormous liberation of energy in the form of light and heat. This makes possible the manufacture of hydrogen bombs, which are virtually unlimited in size and destructive power. Ordinary atomic bombs which work by the disintegration of uranium, have a natural limitation: if too much uranium is collected into one place, it is liable to explode by itself, without waiting to be detonated, so that uranium bombs cannot be made with more than a certain maximum size. But a hydrogen bomb may contain as much hydrogen as we please, because hydrogen by itself is not explosive. It is only when the hydrogen is detonated by a conventional uranium bomb that it combines to form helium and release energy. This is because the combination can only take place at a very high temperature.

There is a further advantage: the supply of uranium in the planet is very limited, and it might be feared that it would be used up before the human race were exterminated, but now that the practically unlimited supply of hydrogen can be utilized, there is considerable reason to hope that *homo sapiens* may put an end to himself, to the great advantage of such less ferocious animals as may survive.

But it is time to return to less cheerful topics.

We have thus two kinds of mass, neither of which quite fulfils the old ideal. The mass as measured by an observer who is in motion relative to the body in question is a relative quantity, and has no physical significance as a property of the body.

The 'proper mass' is a genuine property of the body, not dependent upon the observer; but it, also, is not strictly constant. As we shall see shortly, the notion of mass becomes absorbed into the notion of energy; it represents, so to speak, the energy which the body expends internally, as opposed to that which it displays to the outer world.

Conservation of mass, conservation of momentum, and conservation of energy were the great principles of classical mechanics. Let us next consider conservation of momentum.

The momentum of a body in a given direction is its velocity in that direction multiplied by its mass. Thus a heavy body moving slowly may have the same momentum as a light body moving fast. When a number of bodies interact in any way, for instance by collisions, or by mutual gravitation, so long as no outside influences come in, the total momentum of all the bodies in any direction remains unchanged. This law remains true in the theory of relativity. For different observers, the mass will be different, but so will the velocity; these two differences neutralize each other, and it turns out that the principle still remains true.

The momentum of a body is different in different directions The ordinary way of measuring it is to take the velocity in a given direction (as measured by the observer) and multiply it by the mass (as measured by the observer). Now the velocity in a given direction is the distance travelled in that direction in unit time. Suppose we take instead the distance travelled in that direction while the body moves through unit 'interval.' (In ordinary cases, this is only a very slight change, because, for velocities considerably less than that of light, interval is very nearly equal to lapse of time.) And suppose that instead of the mass as measured by the observer we take the proper mass. These two changes increase the velocity and diminish the mass, both in the same proportion. Thus the momentum remains the same, but the quantities that vary according to the observer have been replaced by quantities which are fixed independently of the observer—with the exception of the distance travelled by the body in the given direction.

When we substitute space-time for time, we find that the

measured mass (as opposed to the proper mass) is a quantity of the same kind as the momentum in a given direction; it might be called the momentum in the time-direction. The measured mass is obtained by multiplying the invariant mass by the *time* traversed in travelling through unit interval; the momentum is obtained by multiplying the same invariant mass by the *distance* traversed (in the given direction) in travelling through unit interval. From a space-time point of view, these naturally belong together.

Although the measured mass of a body depends upon the way the observer is moving relatively to the body, it is none the less a very important quantity. The conservation of measured mass is the same thing as the conservation of energy. This may seem surprising, since at first sight mass and energy are very different things. But it has turned out that energy is the same thing as measured mass. To explain how this comes about is not easy: nevertheless we will make the attempt.

In popular talk, 'mass' and 'energy' do not mean at all the same thing. We associate 'mass' with the idea of a fat man in a chair, very slow to move, while 'energy' suggests a thin person full of hustle and 'pep.' Popular talk associates 'mass' with 'inertia,' but its view of inertia is one-sided: it includes slowness in beginning to move, but not slowness in stopping, which is equally involved. All these terms have technical meanings in physics, which are only more or less analogous to the meanings of the terms in popular talk. For the present, we are concerned with the technical meaning of 'energy.'

Throughout the latter half of the nineteenth century, a great deal was made of the 'conservation of energy,' or the 'persistence of force,' as Herbert Spencer preferred to call it. This principle was not easy to state in a simple way, because of the different forms of energy; but the essential point was that energy is never created or destroyed, though it can be transformed from one kind into another. The principle acquired its position through Joule's discovery of the 'mechanical equivalent of heat,' which showed that there was a constant proportion between the work required to produce a given amount of heat and the work required to raise a given weight through a given height: in fact, the same sort of work could be utilized for either purpose according to the mechanism. When heat was found to consist of motion in molecules, it was seen to be natural that it should be analogous to other forms of energy. Broadly speaking, by the help of a certain amount of theory, all forms of energy were reduced to two, which were called respectively 'kinetic' and 'potential.' These were defined as follows:

The kinetic energy of a particle is half the mass multiplied by the square of the velocity. The kinetic energy of a number of particles is the sum of the kinetic energies of the separate particles.

The potential energy is more difficult to define. It represents any state of strain, which can only be preserved by the application of force. To take the easiest case: if a weight is lifted to a height and kept suspended, it has potential energy, because, if left to itself, it will fall. Its potential energy is equal to the kinetic energy which it would acquire in falling through the same distance through which it was lifted. Similarly, when a comet goes round the sun in a very eccentric orbit, it moves much faster when it is near the sun than when it is far from it, so that its kinetic energy is much greater when it is near the sun. On the other hand its potential energy is greatest when it is farthest from the sun, because it is then like the stone which has been lifted to a height. The sum of the kinetic and potential energies of the comet are constant, unless it suffers collisions or loses some of its material. We can determine accurately the *change* of potential energy in passing from one position to another, but the total amount of it is to a certain extent arbitrary, since we can fix the zero level where we like. For example, the potential energy of our stone may be taken to be the kinetic energy it would acquire in falling to the surface of the earth, or what it would acquire in falling down a well to the centre of the earth, or any assigned lesser distance. It does not matter which we take, so long as we stick to our decision. We are concerned with a profit-and-loss account, which is unaffected by the amount of the assets with which we start.

Both the kinetic and the potential energies of a given set

of bodies will be different for different observers. In classical dynamics, the kinetic energy differed according to the state of motion of the observer, but only by a constant amount; the potential energy did not differ at all. Consequently, for each observer, the total energy was constant—assuming always that the observers concerned were moving in straight lines with uniform velocities, or, if not, were able to refer their motions to bodies which were so moving. But in relativity dynamics the matter becomes more complicated. The Newtonian ideas of kinetic and potential energy can without much difficulty be adapted to the special theory of relativity. But we cannot profitably adapt the idea of potential energy to the general theory of relativity, nor can we generalize the idea of kinetic energy, except in the case of a single body. Therefore the conservation of energy, in the usual Newtonian sense, cannot be maintained. The reason is that the kinetic and potential energies of a system of bodies are inherently ideas which refer to extended regions of space-time. The very wide latitude in choice of co-ordinates, and the hilly character of space-time, which were explained in Chapter VIII, combine to make it very awkward to introduce ideas of this sort into the general theory. There is a conservation law in the general theory, but it is not as useful as the conservation laws in Newtonian mechanics and in the special theory, because it depends on the choice of co-ordinates in a way which is difficult to understand. We have seen that independence of the choice of co-ordinates is a guiding principle in the general theory of relativity, and the conservation law is suspect because it conflicts with this principle. Whether this means that conservation is of lesser fundamental importance than was thought hitherto, or whether a satisfactory conservation law still lies hidden in the mathematical complexities of the theory, is a question which has still to be resolved. In the meantime, we must in the general theory be satisfied with the idea of kinetic energy for a single particle only. That is all we shall need in the argument which follows. It should be remembered that these difficulties about the conservation of energy arise only in the general theory, and not in the special theory. Whenever gravitation may be neglected and the special theory becomes applicable, the conservation of energy can be maintained.

What is meant by 'conservation' in practice is not exactly what it means in theory. In theory we say that a quantity is conserved when the amount of it in the world is the same at any one time as at any other. But in practice we cannot survey the whole world, so we have to mean something more manageable. We mean that, taking any given region, if the amount of the quantity in the region has changed, it is because some of the quantity has passed across the boundary of the region. If there were no births and deaths, population would be conserved; in that case the population of a country could only change by emigration or immigration, that is to say, by passing across the boundaries. We might be unable to take an accurate census of China or Central Africa, and therefore we might not be able to ascertain the total population of the world. But we should be justified in assuming it to be constant if, wherever statistics were possible, the population never changed except through people crossing the frontiers. In fact, of course, population is not conserved. A physiologist of my acquaintance once put four mice into a thermos. Some hours later, when he went to take them out, there were eleven of them. But mass is not subject to these fluctuations: the mass of the eleven mice at the end of the time was no greater than the mass of the four at the beginning.

This brings us back to the problem for the sake of which we have been discussing energy. We stated that, in relativity theory, measured mass and energy are regarded as the same thing, and we undertook to explain why. It is now time to embark upon this explanation. But here, as at the end of Chapter VI, the totally unmathematical reader will do well to skip, and begin at the following paragraph.

Let us take the velocity of light as the unit of velocity; this is always convenient in relativity theory. Let m be the proper mass of a particle, v its velocity relative to the observer. Then its measured mass will be

$$\frac{m}{\sqrt{1 - v^2}}$$

while its kinetic energy, according to the usual formula, will be

$$\tfrac{1}{2}\, m\, v^2$$

As we saw before, energy only occurs in a profit-and-loss account, so that we can add any constant quantity to it that we like. We may therefore take the energy to be

$$m + \tfrac{1}{2}\, m\, v^2$$

Now if v is a small fraction of the velocity of light, $m + \frac{1}{2} mv^2$ is almost exactly equal to $\frac{m}{\sqrt{1 - v^2}}$. Consequently, for velocities such as large bodies have, the energy and the measured mass turn out to be indistinguishable within the limits of accuracy attainable. In fact, it is better to alter our definition of energy, and take it to be $\frac{m}{\sqrt{1 - v^2}}$, because this is the quantity for which the law analogous to conservation holds. And when the velocity is very great, it gives a better measure of energy than the traditional formula. The traditional formula must therefore be regarded as an approximation, of which the new formula gives the exact version. In this way, energy and measured mass become identified.

I come now to the notion of 'action,' which is less familiar to the general public than energy, but has become more important in relativity physics, as well as in the quantum theory. (The quantum is a small amount of action.) The word 'action' is used to denote energy multiplied by time. That is to say, if there is one unit of energy in a system, it will exert one unit of action in a second, 100 units of action in 100 seconds, and so on; a system which has 100 units of energy will exert 100 units of action in a second, and 10,000 in 100 seconds, and so on. 'Action' is thus, in a loose sense, a measure of how much has been accomplished: it is increased both by displaying more energy and by working for a longer time. Since energy is the same thing as measured mass, we may also take action to be measured mass multiplied by time. In classical mechanics, the 'density' of matter in any region is the mass divided by the volume; that is to say, if you know the density in a small region, you discover the total amount of matter by multiplying the density by the volume of the small region. In relativity mechanics, we always want to substitute space-time for space; therefore a 'region' must no longer be taken to be merely a volume, but a volume lasting for a time; a small region will be a small volume lasting for a small time. It follows that, given the density, a small region in the new sense contains, not a small mass merely, but a small mass multiplied by a small time, that is to say, a small amount of 'action.' This explains why it is to be expected that 'action' will prove of fundamental importance in relativity mechanics. And so in fact it is.

The postulate that a freely-moving particle follows a geodesic may be replaced by an equivalent assumption about the 'action' of the particle. Such an assumption is called a *Principle of Least Action*. This states that, in passing from one state to another, a body chooses a route involving less action than any slightly different route—again a law of cosmic laziness! Principles of least action are not restricted to single bodies. It is possible to make a similar assumption which leads to a description of space-time as a whole, complete with hills and valleys. Such principles, which play a central part in quantum theory as well as in relativity, are the most comprehensive means of stating the purely formal part of mechanics.

CHAPTER XI

The Expanding Universe

WE have been dealing hitherto with experiments and observations most of which concern the earth or the solar system. Only occasionally have we had to reach so far afield as the stars. In this chapter we shall range much farther: we shall see what relativity theory has to say about the universe as a whole.

The astronomical observations which we shall be discussing must be regarded as established scientific results. However, the theoretical explanations of these results are more speculative in character, and it must be supposed that we are dealing with theoretical matters having the same solidity as those with which we have been concerned hitherto. They certainly need improvement. Science does not aim at establishing immutable truths and eternal dogmas: its aim is to approach the truth by successive approximations, without claiming that at any stage final and complete accuracy has been achieved.

A few preliminary explanations about the general appearance of the universe are necessary. Much is now known about the distribution of matter on a very large scale. Our sun is one star in a system of about 100,000 millions of stars called 'the galaxy.' The galaxy is shaped like a giant Catherine wheel, with spiral arms of stars coming out of a bright central hub. The outlines of the galaxy are not very sharp, but the main body of stars is about 100,000 light-years across, and up to one-thirtieth as thick. (A light-year is the distance light travels in a year—about six million million miles.) The sun lies in one of the spiral arms, about 30,000 light-years out from the centre of the hub. The Milky Way, a bright band of stars across the sky which is easily visible on a clear night, is just our edge-on view of the rest of the galaxy from this position in the spiral arm.

Besides stars, the galaxy contains a great deal of gas, mostly hydrogen, and dust. The total mass of the gas and dust is probably about one quarter the total mass of all the stars put together. The whole accumulation of stars, dust and gas rotates slowly around the hub. The speed of rotation varies with distance from the hub: the sun takes about 250 million years to go once right round.

The galaxy is by no means alone in the universe. It is one among many millions of similar systems scattered throughout the region which our telescopes can explore. The other systems are also called galaxies (or sometimes 'nebulae'). Some galaxies are flattened, with spiral arms like our own, others are round like footballs or oval like rugby balls, still others quite irregular in shape.

Galaxies show a distinct tendency to be collected into groups. These groups are called 'clusters.' A single cluster may contain up to a thousand or so galaxies,each of which is a vast star system like our own. Our own galaxy belongs to a cluster, called the 'local group,' which has about seventeen other galaxies in it (we cannot be quite certain how many there are because several of the suspected members are comparatively small and very faint). Our best-known neighbour in the local group is the great Andromeda galaxy, which is about 2,000,000 light-years away. It is faintly visible to the naked eye.

Clusters of galaxies are the largest easily identifiable units of matter in the universe. There is some evidence of grouping into larger units—into clusters of clusters, but this is not certain. Aside from this, the distribution of clusters seems to be fairly uniform. There are about as many in one part of the sky as in another, and they appear to be uniformly distributed in depth. Of course the clusters are not spaced regularly like rows of dots. They are distributed haphazardly, like rainspots on a windowpane just after it has begun to rain. The distribution of clusters is uniform in the same sense that the distribution of rainspots is uniform—you cannot say that the number of rainspots on every windowpane is the same, but the number will not vary much from one pane to the next.

Because the clusters of galaxies are the largest natural units, and because we can already see large numbers of these units, it is reasonable to think that the part of the universe visible through existing telescopes is typical of the universe as a whole. It would not be reasonable to suppose that the uniform region extends just as far as telescopes can see now (which is about 3,500 million light-years) and that the next improvement in observation will discover more distant regions of a quite different character. It would not be impossible for this to be so, but it would mean that the local group, or somewhere near it, is picked out especially as the centre of the uniform region, while there is no scientifiic evidence for supposing it to be picked out in this way.

This idea that the universe is uniform on a large scale, which was suggested long before there was adequate astronomical evidence for it, has now acquired the status of a fundamental postulate. It is usually called the 'cosmological principle.' The cosmological principle is really only an extension of Copernicus's ideas. As soon as we give up the egotistical notion that the earth is at the centre of all things, we are forced to realize that the sun, which is an ordinary star, has no more claim than the earth to a special place in our description of the universe. When we find that our galaxy and the cluster to which it belongs are also typical specimens, then they too must be placed logically on a par with other similar objects. Nor is there any empirical reason for supposing that the laws of physics vary systematically from one cluster of galaxies to the next.

We conclude from such arguments that the universe is uniform on a large scale. In other words, it conforms to the cosmological principle.

The implications of this may be put in a slightly different way. Suppose you were put into a box without windows and transported to a distant part of the universe. When released from the box you would not, of course, see the particular distribution of stars and galaxies which is visible from the earth—the geographical details of your new environment would be different—but according to the cosmological principle, the overall appearance of the universe would be the same. Aside from details, you would not be able to tell what part of the universe you were in.

There is one very remarkable phenomenon which might have led us to suppose that our local cluster of galaxies does after all have a special position in the universe. This is the so-called 'red-shift' in the spectra of distant galaxies. As well shall now see, it is because of this phenomenon that the universe is said to be expanding.

We are concerned here with an effect which was explained in Chapter IX,[1] although in that chapter we were not directly concerned with it. You will remember the analogy with sound which was introduced then: if a train is moving towards you then the pitch of its whistle is higher than if it is standing still, while if it is moving away from you the pitch is lower. The effects are very similar in the case of light. If the source of light is moving towards you, then the whole spectrum of the light is shifted towards the violet; if the source is moving away from you, then the whole spectrum is shifted towards the red. These shifts of the spectrum correspond to the changes of pitch of the train whistle. The amount of the shift depends on the speed of the source of light relative to you. (This has nothing to do with the speed of the light itself, which as we have seen is independent of the motion of its source.) This shift of the spectrum provides a means of determining the speeds of stars and galaxies, by comparing the spectra of the light which they send out with similar spectra produced in laboratories on earth. The speeds of galaxies in the local group, measured in this way, range up to about 300 miles a second. This is very fast by everyday standards, but because of the great distances between the galaxies it would be millions of years before there was any noticeable change in their positions.

Some of the galaxies in the local group are moving towards us, others away from us. There is nothing very remarkable about this motion, which might be compared to the motion of bees in a swarm. The bees move about relative to one another, but the swarm as a whole keeps together. The situation is rather different when we come to examine clusters other than our own. Here again there are internal motions in each cluster, but all the other clusters appear to be moving *away* from our

[1] On pages 83 and 84.

own, and the further away they are, the faster they appear to be moving. It is this remarkable phenomenon which suggests that the universe is expanding.

Because all the other clusters appear to be moving away from ours, we might be inclined to think that the local group is in some way at the centre of the expanding universe. This would be a mistake, because it ignores the relative character of motion which has been pointed out repeatedly in this book. Consider again the analogy with swarms of bees. Suppose that they are very well-trained swarms, which hover above the ground ten yards apart in a line running from west to east. Then suppose that one of the swarms stays at rest relative to the ground, while the swarm ten yards to the east of it moves east at a yard a minute, the swarm twenty yards to the east moves easy at two yards a minute, and so on, while the swarms to the west of the fixed swarm move west at similar speeds. Then it will appear to a bee in any of the swarms, fixed or moving, that all the other swarms are moving away from his at speeds proportional to their distances. If the ground were not available as a standard of rest, then there would be no reason to think that any one of the swarms was picked out in a special way.

The behaviour of the clusters of galaxies is entirely similar. Of course they are distributed irregularly in all directions instead of being lined up like our well-trained swarms, but as in the case of the swarms, it appears to an observer in any cluster that all the others are moving away from his. Since there is no absolute standard of rest in the universe, the appearance of expansion is the same for all the clusters.

The nearest cluster, which is about 43 million light-years away, and which contains about 2,500 galaxies, has a red-shift corresponding to a speed of recession from us of 750 miles a second. The most distant cluster so far investigated has a red-shift over one hundred times as great, corresponding to a speed of recession which is almost half of the speed of light. (Speeds of recession corresponding to red-shifts as large as this must be calculated on the basis of the Lorentz transformation formulae given in Chapter VI).

The largest astronomical red-shifts discovered hitherto are not those of distant clusters, but of the so-called 'quasistellar objects' (quasars) whose red-shifts correspond to speeds of recession up to four-fifths of the speed of light. However, the nature of these objects is not yet understood, and so they cannot yet be taken properly into account when the astronomical data are used to construct a theoretical model.

Let us now examine how this information about the universe can be fitted into the general relativity theory. We have seen that the gravitational effects of the sun may be described as those of a hill in space-time. A galaxy or a cluster may be represented in the same way, but by a much larger hill, because of its much greater mass. (The mass of a typical cluster is about a million million times the mass of the sun.) If we tried to incorporate into this description details of the distribution of stars in each galaxy and galaxies in each cluster, we should have a very complicated hill with many peaks and valleys. We could then try to describe the whole universe in a way which could be represented by a space-time with hills, representing the clusters, scattered about in it. Such a description would be mathematically very complicated, because it would include many 'geographical' details not essential to a description of the overall appearance of the universe. In order to simplify the description, we construct models which preserve what seem to be the essential features while leaving out the geographical details. The features which we preserve are the large scale uniformity and the expansion. The details left out are the precise positions, sizes, and compositions of the individual clusters.

Thus we construct model space-times to represent the universe by supposing it to be exactly, instead of approximately, uniform. In these simplified models we imagine matter to be smoothed out into continuous distribution instead of being collected into clusters with large spaces in between them.

Just as the accumulation of matter into a cluster can be described by saying that there is a large hill in space-time where we see the cluster, or by saying that space-time is curved nearby to the cluster, so the uniform distribution of matter in a smoothed-out model of the universe can be described by saying that space-time is curved uniformly. The effect of smoothing out the matter composing the different clusters is to smooth out the corresponding curvature to produce a slight overall curvature. This overall curvature of the universe is somewhat analogous to the curvature of a sphere in ordinary space, but we shall not push the analogy of curvature with space-time hills any further, by comparing the overall curvature of space-time with the curvature of the earth, because this might easily become misleading.

Einstein's law of gravitation, combined with the smoothing-out assumption—the assumption of exact uniformity—allows us to construct a variety of models of the universe, in which the overall curvature takes a variety of different forms. The main effect of this overall curvature is that it implies, in some of the models that the spectra of distant objects will be shifted towards the red. It is largely a matter of taste whether this red-shift is attributed to a recessional motion, or to space-time curvature. The effect will appear in one guise or the other, depending on the coordinate system which is used to describe the universe. What relativity predicts does not, of course, depend on the choice of coordinate system.

The model universes which we have been considering agree more or less well with observations of the overall properties of our own universe. There are others, equally consistent with Einstein's law and with the assumption of uniformity, in which there is a blue-shift, corresponding to a contraction of the universe, instead of a red-shift. The existence of such models is no reason for rejecting Einstein's theory. It implies that the theory is not complete—some additional assumption is required which will exclude the unwanted models. Various assumptions have been suggested, but so far an entirely satisfactory one has not been found.

Let us examine the consequences of expansion a little further, remembering always that what we say may always be rephrased in terms of space-time curvature if that becomes necessary. The most obvious consequences is that if the universe is, so to say, thinning out—if the clusters of galaxies are getting farther and farther apart, then in the past they must have been closer together than they are now. Suppose we were to take a movie film of the expanding universe, over a period of many millions of years, so as to record the whole history of the expansion. If such a film were to be shown backwards, then it would show the history of the universe in reverse. Instead of moving away from one another, all the clusters of galaxies would appear to be moving towards one another. As the film ran back, they would get closer and closer together, until presumably they were so close together that there were no gaps between them any more. Still further back, we may suppose, even the spaces between the stars would be closed up, all the available space being filled up with highly condensed hot gas out of which the stars could have evolved. Recent astronomical observations of short radio waves seem to confirm the existence of this highly condensed state. It seems that a certain proportion of the radio energy arriving at receivers on the earth cannot be attributed to emission by stars or by the interstellar gas, but agrees reasonably well with what might be expected to be visible now of the radiation present in the universe at an early stage when all matter was in a highly condensed state. However, the predictions of theoretical models about this condensed state cannot be trusted too far. What is known of the quantum properties of matter suggests that at a sufficiently early time, these properties would have had important effects. There is no general agreement about exactly when this would have been so, but it seems likely at least that quantum effects could not have been negligible at the stage when the whole of the now observable universe was compressed, say, to the size of a proton. We have seen that Einstein's theory is unable to describe such effects, so that there is in fact no reliable information about the nature of the universe at this stage in its expansion. Besides this, the possibility of quantum effects implies that nothing which occurred before the highly condensed state could possibly influence the subsequent behaviour of the universe. All this is rather speculative; we may conclude from it only that if the universe has in fact evolved from a higly condensed state, then that highly condensed state represents the earliest time about which there is ever likely to be any scientific information.

Whether such a state actually occurred or not is still under dispute: the available astronomical data are not sufficiently precise to decide the question. Those people who think it did occur are inclined to refer to the highly condensed state as 'the beginning of the universe' or 'the time when the universe was created' or something of that kind. These phrases mean no more than 'the earliest time about which there is ever likely to be any scientific information,' and it is better to avoid them, because they carry undesirable metaphysical implications.

There are other models of the universe consistent with Einstein's law of gravitation, in which the highly condensed state does not occur at all. The best known of these is the so-called steady-state model. We have seen that according to the cosmological principle, you cannot tell where you are in the universe. But two astronomers on planets in different galaxies can tell *when* they are—they both, for example, will observe that the universe is thinning out in the course of the expansion, and can agree on the times at which they respectively observe it to have thinned out to any given extent. In the steady-state model, however, you cannot tell *when* you are any better than you can tell *where*. That is to say, it is assumed in the steady-state model that the universe presents the same overall appearance not only to astronomers in different places, but also to astronomers in the same or different places at different times. The division into space and time which seems to be taking place here is not in conflict with relativity—it applies only to astronomers who move with the clusters of galaxies. An astronomer with a substantially different velocity would make a more complicated description of the universe; we naturally prefer to consider those whose descriptions are simplest.

In order that the overall appearance of the universe should not change with time, despite the expansion, it is evidently necessary that as the thinning out of the clusters of galaxies proceeds, new clusters should appear in between to fill the gaps. Where are the new clusters to come from? According to the steady-state theory, matter must appear in intergalactic space at just such a rate as is necessary to cancel the thinning out by expansion. This matter may be supposed initially in the form of hydrogen gas, which subsequently forms into stars, galaxies and clusters. The rate at which the hydrogen is supposed to appear is very small—one atom in a space the size of St. Paul's Cathedral every thousand years—much too small to be excluded by direct observations, yet large enough to compensate for the thinning out by expansion. The process by which the hydrogen appears is often called 'continual creation,' but this is another term carrying metaphysical overtones, and it is better not to use it. It might seem at first sight that such a process is contrary to the laws of conservation of energy which form part of Einstein's theory. When full account is taken of the overall curvature of the universe, however, it turns out that the suggested process is perfectly consistent with relativity. Of course the rate at which the new atoms appear cannot be just anything at all; the new atoms must appear at just the rate which is required to make up for the expansion.

As things stand at present, certain of the model universes predicting expansion from a highly condensed state are the easiest to reconcile with the astronomical data. All of them have defects, of which the most obvious is that they give only a smoothed-out picture which does not account for the size or composition of the galaxies and clusters. The steady-state model can to some extent overcome this defect, but fails to give a satisfactory explanation of certain other data, for example, those which suggest a highly condensed state in the remote past.

The construction of an entirely satisfactory model depends on the resolving of some serious mathematical difficulties; which of the available models is to be preferred at any particular time must depend on the astronomical data.

CHAPTER XII

Conventions and Natural Laws

ONE of the most difficult matters in all controversy is to distinguish disputes about words from disputes about facts: it ought not to be difficult, but in practice it is. This is quite as true in physics as in other subjects. In the seventeenth century there was a terrific debate as to what 'force' is; to us now, it was obviously a debate as to how the word 'force' should be defined, but at the time it was thought to be much more. One of the purposes of the method of tensors, which is employed in the mathematics of relativity, is to eliminate what is purely verbal (in an extended sense) in physical laws. It is of course obvious that what depends on the choice of co-ordinates is 'verbal' in the sense concerned. A man punting walks along the boat, but keeps a constant position with reference to the river-bed so long as he does not pick up his pole. The Lilliputians might debate endlessly whether he is walking or standing still; the debate would be as to words, not as to facts. If we choose co-ordinates fixed relatively to the boat, he is walking; if we choose co-ordinates fixed relatively to the river-bed, he is standing still. We want to express physical laws in such a way that it shall be obvious when we are expressing the same law by reference to two different systems of co-ordinates, so that we shall not be misled into supposing we have different laws when we only have one law in different words. This is accomplished by the method of tensors. Some laws which seem plausible in one language cannot be translated into another; these are impossible as laws of nature. The laws that can be translated into *any* co-ordinate language have certain characteristics: this is a substantial help in looking for such laws of nature as the theory of relativity can admit to be possible. Of the possible laws, we choose the simplest one which predicts the actual motion of bodies correctly: logic and experience combine in equal proportions in obtaining this expression.

But the problem of arriving at genuine laws of nature is not to be solved by the method of tensors alone; a good deal of careful thought is wanted in addition. Some of this has been done, especially by Eddington; much remains to be done.

To take a simple illustration: suppose, as in the hypothesis of the Fitzgerald contraction, that lengths in one direction were shorter than in another. Let us assume that a foot-rule pointing north is only half as long as the same foot-rule pointing east, and that this is equally true of all other bodies. Does such a hypothesis have any meaning? If you have a fishing-rod fifteen feet long when it is pointing west, and you then turn it to the north, it will still measure fifteen feet, because your foot-rule will have shrunk too. It won't 'look' any shorter, because your eye will have been affected in the same way. If you are to find out the change, it cannot be by ordinary measurement: it must be by some such method as the Michelson–Morley experiment, in which the velocity of light is used to measure lengths. Then you still have to decide whether it is simpler to suppose a change of length or a change in the velocity of light. The experimental fact would be that light takes longer to traverse what your foot-rule declares to be a given distance in one direction than in another—or, as in the Michelson–Morley experiment, that it ought to take longer but doesn't. You can adjust your measures to such a fact in various ways; in any way you choose to adopt, there will be an element of convention. This element of convention survives in the laws that you arrive at after you have made your decision as to measures, and often it takes subtle and elusive forms. To eliminate the element of convention is, in fact, extraordinarily difficult; the more the subject is studied, the greater the difficulty is seen to be.

A more important example is the question of the size of the electron. We find experimentally that all electrons are the same size. How far is this a genuine fact ascertained by experiment, and how far is it a result of our conventions of measurement? We have here two different comparisons to make: (1) in regard to one electron at different times; (2) in regard to two electrons at the same time. We can then arrive at the comparison of two electrons at different times, by combining (1) and (2). We may dismiss any hypothesis which would affect all electrons equally; for example, it would be useless to suppose that in one region of space-time they were all larger than in another. Such a change would affect our measuring appliances just as much as the things measured, and would therefore produce no discoverable phenomena. This is as much as to say that it would be no change at all. But the fact that two electrons have the same mass, for instance, cannot be regarded as purely conventional. Given sufficient minuteness and accuracy, we could compare the effects of two different electrons upon a third; if they were equal under like circumstances, we should be able to infer equality in a not purely conventional sense.

Eddington describes the process concerned in the more advanced portions of the theory of relativity as 'world-building.' The structure to be built is the physical world as we know it; the economical architect tries to construct it with the smallest possible amount of material. This is a question for logic and mathematics. The greater our technical skill in these two subjects, the more real building we shall do, and the less we shall be content with mere heaps of stones. But before we can use in our building the stones that nature provides, we have to hew them into the right shapes: this is all part of the process of building. In order that this may be possible, the raw material must have *some* structure (which we may conceive as analogous to the grain in timber), but almost any structure will do. By successive mathematical refinements, we whittle away our initial requirements until they amount to very little. Given this necessary minimum of structure in the raw material, we find that we can construct from it a mathematical expression which will have the properties that are needed for describing the world we perceive—in particular, the properties of conservation which are characteristic of momentum and energy (or mass). Our raw material consisted merely of events; but when we find that we can build out of it something which, as measured, will seem to be never created or destroyed, it is not surprising that we should come to believe in 'bodies.' These are really mere mathematical constructions out of events, but owing to their permanence they are practically important, and our senses (which were presumably developed by biological needs) are adapted for noticing them, rather than the crude continuum of events which is theoretically more fundamental. From this point of view, it is astonishing how little of the real world is revealed by physical science: our knowledge is limited, not only by the conventional element, but also by the selectiveness of our perceptual apparatus.

In particular, conditions of symmetry may be entirely created by conventions as to measurement, and there is no reason to suppose that they represent any property of the real world. The law of gravitation itself, according to Eddington, may be regarded as expressing conventions of measurement. 'The conventions of measurement,' he says, 'introduce an isotropy[1] and homogeneity into measured space which need not originally have any counterpart in the relation-structure which is being surveyed. This isotropy and homogeneity is exactly expressed by Einstein's law of gravitation.'[2]

The limitations of knowledge introduced by the selectiveness of our perceptual apparatus may be illustrated by the indestructibility of energy. This has been gradually discovered by experiment, and seemed a well-founded empirical law of nature. Now it turns out that, from our original space-time continuum we can construct a mathematical expression which will have properties causing it to appear indestructible. The statement that energy is indestructible then ceases to be a proposition of physics, and becomes instead a proposition of linguistics and psychology. As a proposition of linguistics: 'Energy' is the name of the mathematical expression in question. As a proposition of psychology: our senses are such that we notice what is roughly the mathematical expression in question, and we are led nearer and nearer to it as we refine upon our crude perceptions by scientific observation. This is much less than physicists used to think they knew about energy.

[1] 'Isotropy' means being similar in all directions—e.g., that a foot-rule is as long when it points north as when it points east.

[2] *Mathematical Theory of Relativity*, p. 238.

The reader may say: What then is left of physics? What do we really know about the world of matter? Here we may distinguish three departments of physics. There is first what is included within the theory of relativity, generalized as widely as possible. Next, there are laws which cannot be brought within the scope of relativity. Thirdly, there is what may be called geography. Let us consider each of these in turn.

The theory of relativity, apart from convention, tells us that the events in the universe have a four-dimensional order, and that, between any two events which are near together in this order, there is a relation called 'interval,' which is capable of being measured if suitable precautions are taken. It tells us also that 'absolute motion,' 'absolute space,' and 'absolute time' cannot have any physical significance; laws of physics involving these concepts are not acceptable. This is hardly a physical law in itself, but rather a useful rule to enable us to reject some proposed physical laws as unsatisfactory.

Beyond this, there is little in the theory of relativity that can be regarded as physical laws. There is a great deal of mathematics, showing that certain mathematically-constructed quantities must behave like the things we perceive; and there is a suggestion of a bridge between psychology and physics in the theory that these mathematically-constructed quantities are what our senses are adapted for perceiving. But neither of these things is physics in the strict sense.

The part of physics which cannot, at present, be brought within the scope of relativity is large and important. There is nothing in relativity to show why there should be electrons and protons; relativity cannot give any reason why matter should exist in little lumps. This is the province of the quantum theory, which accounts for many of the properties of matter on the small scale. The quantum theory has been made consistent with the special theory of relativity, but hitherto all attempts to perform a synthesis of quantum theory and general relativity have been unsuccessful. There seem to be very severe difficulties in the way of bringing this part of physics within the framework of general relativity. At present there are equally severe difficulties in the quantum theory itself, and many physicists think that a synthesis of quantum theory and general relativity might solve some of these difficulties. The present situation, as we have seen, is that general relativity accounts fairly satisfactorily for the properties of matter on a very large scale, while quantum theory accounts fairly satisfactorily for the properties of matter on a very small scale. However, there is no apparent connection between the two theories except for their common ground in special relativity theory. This situation is not satisfactory and is unlikely to be permanent. A few people think that general relativity could be extended in such a way as to explain all the results that quantum theory explains, but in a more satisfactory way than present quantum theory does. Einstein towards the end of his life was one of the people who thought this. However, most physicists nowadays think that view is mistaken.

General relativity is the most extreme example of what may be called next-to-next methods. Gravitation need no longer be regarded as due to the effect of the sun on a planet, but may be thought of as expressing the characteristics of the region in which the planet happens to be. These characteristics are supposed to alter bit by bit, gradually, continuously, and not by sudden jumps, as one moves from one part of space-time to another. The effects of electromagnetism may be regarded in a similar way, but as soon as electromagnetism is made to accord with the quantum theory its character changes entirely. The continuous aspect disappears completely, and is replaced by the discontinuous behaviour which as we have already seen is typical of quantum theory. However, if we try to apply to gravitation these ideas of quantum theory we find that they do not fit properly, and that some considerable alteration in one theory or the other, or both, is necessary. What modification is needed we do not yet know.

The difficulty may be explained in a somewhat different way. When an astronomer observes the sun, the sun preserves a lordly indifference to his proceedings. But when a physicist tries to find out what is happening to an atom, the apparatus which he uses is much larger than the thing he is observing, instead of much smaller, and is likely to have some effect on it.

It is found that the sort of apparatus best suited for determining the position of an atom is bound to affect its velocity, and the sort of apparatus best suited for determining the velocity is bound to affect its position. This does not cause any difficulty when the quantum theory of atoms is made to accord with the special theory of relativity, because then gravitation is neglected, and the space-time is supposed to be flat whether there are atoms about in it or not. But if we try to make quantum theory accord with the general theory of relativity, then gravitation is not to be neglected, so that the curvature of space-time will depend on the whereabouts of the atoms. However, as we have just seen, the quantum theory makes it quite clear that we cannot always know where the atoms are. This is the root of the difficulty.

Finally we come to geography, in which I include history. The separation of history from geography rests upon the separation of time from space: when we amalgamate the two in space-time, we need one word to describe the combination of geography and history. For the sake of simplicity, I shall use the one word geography in this extended sense.

Geography, in this sense, includes everything that, as a matter of crude fact, distinguishes one part of space-time from another. One part is occupied by the sun, one by the earth; the intermediate regions contain light-waves, but no matter (apart from a very little here and there). There is a certain degree of theoretical connection between different geographical facts; to establish this is the purpose of physical laws.

We are already in a position to calculate the large facts about the solar system backwards and forwards for vast periods of time. But in all such calculations we need a basis of crude fact. The facts are interconnected, but facts can only be inferred from other facts, not from general laws alone. Thus the facts of geography have a certain independent status in physics. No amount of physical laws will enable us to infer a physical fact unless we know other facts as data for our inference. And here when I speak of 'facts' I am thinking of particular facts of geography, in the extended sense in which I am using the term.

In the theory of relativity, we are concerned with *structure*, not with the material of which the structure is composed. In geography, on the other hand, the material is relevant. If there is to be any difference between one place and another, there must either be differences between the material in one place and that in another, or places where there is material and places where there is none. The former of these alternatives seems the more satisfactory. We might try to say: There are electrons and protons and the other sub-atomic particles, and the rest is empty. But in the empty regions there are light-waves, so that we cannot say that there is nothing there. According to quantum theory, we cannot even say exactly where things are, but only that one place is more likely than another to find an electron in. Some people maintain that light-waves, and particles as well, are just disturbances in the aether, others are content to say that they are just disturbances; but in any case events are occurring wherever there are likely to be light-waves or particles. That is all that we can say for the places where there is likely to be energy in one form or another, since energy has turned out to be a mathematical construction built out of events. We may say, therefore, that there are events everywhere in space-time, but they must be of a somewhat different kind according as we are dealing with a region where there is very likely to be an electron or proton, or with the sort of region we should ordinarily call empty. But as to the intrinsic nature of these events we can know nothing, except when they happen to be events in our own lives. Our own perceptions and feelings must be part of the crude material of events which physics arranges into a pattern—or rather, which physics finds to be arranged in a pattern. As regards events which do not form part of our own lives, physics tells us the pattern of them, but is quite unable to tell us what they are like in themselves. Nor does it seem possible that this should be discovered by any other method.

CHAPTER XIII

The Abolition of 'Force'

IN the Newtonian system, bodies under the action of no forces move in straight lines with uniform velocity; when bodies do not move in this way, their change of motion is ascribed to a 'force.' Some forces seem intelligible to our imagination: those exerted by a rope or string, by bodies colliding, or by any kind of obvious pushing or pulling. As explained in an earlier chapter, our apparent imaginative understanding of these processes is quite fallacious; all that it really means is that past experience enables us to foresee more or less what is going to happen without the need of mathematical calculations. But the 'forces' involved in gravitation and in the less familiar forms of electrical action do not seem in this way 'natural' to our imagination. It seems odd that the earth can float in the void: the natural thing to suppose is that it must fall. That is why it has to be supported on an elephant, and the elephant on a tortoise, according to some early speculators. The Newtonian theory, in addition to action at a distance, introduced two other imaginative novelties. The first was, that gravitation is not always and essentially directed what we should call 'downwards,' i.e. towards the centre of the earth. The second was, that a body going round and round in a circle with uniform velocity is not 'moving uniformly' in the sense in which that phrase is applied to the motion of bodies under no forces, but is perpetually being turned out of the straight course towards the centre of the circle, which requires a force pulling it in that direction. Hence Newton arrived at the view that the planets are attracted to the sun by a force, which is called gravitation.

This whole point of view, as we have seen, is superseded by relativity. There are no longer such things as 'straight lines' in the old geometrical sense. There are 'straightest lines,' or geodesics, but these involve time as well as space. A light-ray passing through the solar system does not describe the same orbit as a comet, from a geometrical point of view; nevertheless each moves in a geodesic. The whole imaginative picture is changed. A poet might say that water runs downhill because it is attracted to the sea, but a physicist or an ordinary mortal would say that it moves as it does, at each point, because of the nature of the ground at that point, without regard to what lies ahead of it. Just as the sea does not cause the water to run towards it, so the sun does not cause the planets to move round it. The planets move round the sun because that is the easiest thing to do—in the technical sense of 'least action.' It is the easiest thing to do because of the nature of the region in which they are, not because of an influence emanating from the sun.

The supposed necessity of attributing gravitation to a 'force' attracting the planets towards the sun has arisen from the determination to preserve Euclidean geometry at all costs. If we suppose that our space is Euclidean, when in fact it is not, we shall have to call in physics to rectify the errors of our geometry. We shall find bodies not moving in what we insist upon regarding as straight lines, and we shall demand a cause for this behaviour. Eddington has stated this matter with admirable lucidity. He supposes a physicist who has assumed the formula for interval which is used in the special theory of relativity—a formula which still supposes that the observer's space is Euclidean. He continues:

'Since intervals can be compared by experimental methods, he ought soon to discover that his (formula for the interval) cannot be reconciled with observational results, and so realize his mistake. But the mind does not so readily get rid of an obsession. It is more likely that our observer will continue in his opinion, and attribute the discrepancy of the observations to some influence which is present and affects the behaviour of his test-bodies. He will, so to speak, introduce a supernatural agency which he can blame for the consequences of his mistake. . . . The name given to any agency which causes deviation from uniform motion in a straight line is *force* according to the Newtonian definition of force. Hence the agency invoked through our observer's mistake is described as a "field of force". . . .

A field of force represents the discrepancy between the natural geometry of a co-ordinate system and the abstract geometry arbitrarily ascribed to it.'[1]

If people were to learn to conceive the world in the new way, without the old notion of 'force,' it would alter not only their physical imagination, but probably also their morals and politics. The latter effect would be quite illogical, but is none the less probable on that account. In Newton's theory of the solar system, the sun seems like a monarch whose behests the planets have to obey. In Einstein's world there is more individualism and less government than in Newton's. There is also far less hustle: we have seen that laziness is the fundamental law of Einstein's universe. The word 'dynamic' has come to mean, in newspaper language, 'energetic and forceful'; but if it meant 'illustrating the principles of dynamics,' it ought to be applied to the people in hot climates who sit under banana trees waiting for the fruit to drop into their mouths. I hope that journalists, in future, when they speak of a 'dynamic personality,' will mean a person who does what is least trouble at the moment, without thinking of remote consequences. If I can contribute to this result, I shall not have written in vain.

It has been customary for people to draw arguments from the laws of nature as to what we ought to do. Such arguments seem to me a mistake: to imitate nature may be merely slavish. But if nature, as portrayed by Einstein, is to be our model, it would seem that the anarchists will have the best of the argument. The physical universe is orderly, not because there is a central government, but because every body minds its own business. No two particles of matter ever come into contact. When they get too close, they both move off. If a man were had up for knocking another man down, he would be scientifically correct in pleading that he had never touched him. What happened was that there was a hill in space-time in the region of the other man's nose, and it fell down the hill.

The abolition of 'force' seems to be connected with the substitution of sight for touch as the source of physical ideas,

[1] *Mathematical Theory of Relativity*, pp. 37–8. Italics in the original.

as explained in Chapter I. When an image in a looking-glass moves, we do not think that something has pushed it. In places where there are two large mirrors opposite to each other, you may see innumerable reflections of the same object. Suppose a gentleman in a top-hat is standing between the mirrors, there may be twenty or thirty top-hats in the reflections. Suppose now somebody comes and knocks off the gentleman's hat with a stick: all the other twenty or thirty top-hats will tumble down at the same moment. We think that a force is needed to knock off the 'real' top-hat, but we think the remaining twenty or thirty tumble off, so to speak, of themselves, or out of a mere passion for imitation. Let us try to think out this matter a little more seriously.

Obviously something happens when an image in a looking-glass moves. From the point of view of sight, the event seems just as real as if it were not in a mirror. But nothing has happened from the point of view of touch or hearing. When the 'real' top-hat falls, it makes a noise; the twenty or thirty reflections fall without a sound. If it falls on your toe, you feel it; but we believe that the twenty or thirty people in the mirrors feel nothing, though top-hats fall on their toes too. But all this is equally true of the astronomical world. It makes no noise, because sound cannot travel across a vacuum. So far as we know, it causes no 'feelings,' because there is no one on the spot to 'feel' it. The astronomical world, therefore, seems hardly more 'real' or 'solid' than the world in the looking-glass, and has just as little need of 'force' to make it move.

The reader may feel that I am indulging in idle sophistry. 'After all,' he may say, 'the image in the mirror is the reflection of something solid, and the top-hat in the mirror only falls off because of the force applied to the real top-hat. The top-hat in the mirror cannot indulge in behaviour of its own; it has to copy the real one. This shows how different the image is from the sun and the planets, because *they* are not obliged to be perpetually imitating a prototype. So you had better give up pretending that an image is just as real as one of the heavenly bodies.'

There is, of course, some truth in this; the point is to discover exactly *what* truth. In the first place, images are not 'imaginary.' When you see an image, certain perfectly real light-waves reach your eye; and if you hang a cloth over the mirror, these light-waves cease to exist. There is, however, a purely optical difference between an 'image' and a 'real' thing. The optical difference is bound up with this question of imitation. When you hang a cloth over the mirror, it makes no difference to the 'real' object; but when you move the 'real' object away, the image vanishes also. This makes us say that the light-rays which make the image are only reflected at the surface of the mirror, and do not really come from a point behind it, but from the 'real' object. We have here an example of a general principle of great importance. Most of the events in the world are not isolated occurrences, but members of groups of more or less similar events, which are such that each group is connected in an assignable manner with a certain small region of space-time. This is the case with the light-rays which make us see both the object and its reflection in the mirror: they all emanate from the object as a centre. If you put an opaque globe round the object at a certain distance, the object and its reflection are invisible at any point outside the globe. We have seen that gravitation, although no longer regarded as an action at a distance, is still connected with a centre: there is, so to speak, a hill symmetrically arranged about its summit, and the summit is the place where we conceive the body to be which is connected with the gravitational field we are considering. For simplicity, common sense lumps together all the events which form one group in the above sense. When two people see the same object, two different events occur, but they are events belonging to one group and connected with the same centre. Just the same applies when two people (as we say) hear the same noise. And so the reflection in a mirror is less 'real' than the object reflected, even from an optical point of view, because light-rays do not spread in *all* directions from the place where the image seems to be, but only in directions in front of the mirror, and only so long as the object reflected remains in position. This illustrates the usefulness of grouping connected events about a centre in the way we have been considering.

When we examine the changes in such a group of objects, we find that they are of two kinds: there are those which affect only some member of the group, and those which make connected alterations in all the members of the group. If you put a candle in front of a mirror, and then hang black cloth over the mirror, you alter only the reflection of the candle as seen from various places. If you shut your eyes, you alter its appearance to you, but not its appearance elsewhere. If you put a red globe round it at a distance of a foot, you alter its appearance at any distance greater than a foot, but not at any distance less than a foot. In all these cases, you do not regard the candle itself as having changed; in fact, in all of them, you find that there are groups of changes connected with a different centre or with a number of different centres. When you shut your eyes, for instance, your eyes, not the candle, look different to any other observer: the centre of the changes that occur is in your eyes. But when you blow out the candle, its appearance *everywhere* is changed; in this case you say that the change has happened to the candle. The changes that happen to an object are those that affect the whole group of events which centre about the object. All this is only an interpretation of common sense, and an attempt to explain what we mean by saying that the image of the candle in the mirror is less 'real' than the candle. There is no connected group of events situated all round the place where the image seems to be, and changes in the image centre about the candle, not about a point about the mirror. This gives a perfectly verifiable meaning to the statement that the image is 'only' a reflection. And at the same time it enables us to regard the heavenly bodies, although we can only see and not touch them, as more 'real' than an image in a looking-glass.

We can now begin to interpret the common-sense notion of one body having an 'effect' upon another, which we must do if we are really to understand what is meant by the abolition of 'force.' Suppose you come into a dark room and switch on the electric light: the appearance of everything in the room is changed. Since everything in the room is visible because it reflects the electric light, this case is really analogous to that of the image in the mirror; the electric light is the centre from which all the changes emanate. In this case, the 'effect' is explained by what we have already said. The more important case is when the effect is a movement. Suppose you let loose a tiger in the middle of a Bank Holiday crowd: they would all move, and the tiger would be the centre of their various movements. A person who could see the people but not the tiger would infer that there was something repulsive at that point. We say in this case that the tiger has an effect upon the people, and we might describe the tiger's action upon them as of the nature of a repulsive force. We know, however, that they fly because of something which happens to *them*, not merely because the tiger is where he is. They fly because they can see and hear him, that is to say, because certain waves reach their eyes and ears. If these waves could be made to reach them without there being any tiger, they would fly just as fast, because the neighbourhood would seem to them just as unpleasant.

Let us now apply similar considerations to the sun's gravitation. The 'force' exerted by the sun only differs from that exerted by the tiger in being attractive instead of repulsive. Instead of acting through waves of light or sound, the sun acquires its apparent power through the fact that there are modifications of space-time all round the sun. Like the noise of the tiger, they are more intense near their source; as we travel away they grow less and less. To say that the sun 'causes' these modifications of space-time is to add nothing to our knowledge. What we know is that the modifications proceed according to a certain rule, and that they are grouped symmetrically about the sun as centre. The language of cause and effect adds only a number of quite irrelevant imaginings, connected with will, muscular tension, and such matters. What we can more or less ascertain is merely the formula according to which space-time is modified by the presence of gravitating matter. More correctly: we can ascertain what kind of space-time *is* the presence of gravitating matter. When space-time is not accurately Euclidean in a certain region, but has a non-Euclidean character which grows more and more marked as we approach a certain centre, and when, further, the departure from Euclid obeys a certain law, we describe this state of affairs briefly by saying that there is gravitating matter at the centre. But this is only a compendious account of what we know. What we know is about the places where the gravitating matter is *not*, not about the place where it is. The language of cause and effect (of which 'force' is a particular case) is thus merely a convenient shorthand for certain purposes; it does not represent anything that is genuinely to be found in the physical world.

And how about matter? Is matter also no more than a convenient shorthand? This question, however, being a large one, demands a separate chapter.

CHAPTER XIV

What is Matter?

THE question 'What is matter?' is of the kind that is asked by metaphysicians, and answered in vast books of incredible obscurity. But I am not asking the question as a metaphysician: I am asking it as a person who wants to find out what is the moral of modern physics, and more especially of the theory of relativity. It is obvious from what we have learned of that theory that matter cannot be conceived quite as it used to be. I think we can now say more or less what the new conception must be.

There were two traditional conceptions of matter, both of which have had advocates ever since scientific speculation began. There were the atomists, who thought that matter consisted of tiny lumps which could never be divided; these were supposed to hit each other and then bounce off in various ways. After Newton, they were no longer supposed actually to come into contact with each other, but to attract and repel each other, and move in orbits round each other. Then there were those who thought that there is matter of some kind everywhere, and that a true vacuum is impossible. Descartes held this view, and attributed the motions of the planets to vortices in the aether. The Newtonian theory of gravitation caused the view that there is matter everywhere to fall into discredit, the more so as light was thought by Newton and his disciples to be due to actual particles travelling from the source of the light. But when this view of light was disproved, and it was shown that light consisted of waves, the aether was revived so that there should be something to undulate. The aether became still more respectable when it was found to play the same part in electromagnetic phenomena as in the propagation of light. It was even hoped that atoms might turn out to be a mode of motion of the aether. At this stage, the atomic view of matter was, on the whole, getting the worst of it.

Leaving relativity aside for the moment, modern physics has provided proof of the atomic structure of ordinary matter, while not disproving the arguments in favour of the aether, to which no such structure is attributed. The result was a sort of compromise between the two views, the one applying to what was called 'gross' matter, the other to the aether. There can be no doubt about electrons and protons, though, as we shall see shortly, they need not be conceived as atoms were conceived traditionally. The truth is, I think, that relativity demands the abandonment of the old conception of 'matter,' which is infected by the metaphysics associated with 'substance,' and represents a point of view not really necessary in dealing with phenomena. This is what we must now investigate.

In the old view, a piece of matter was something which survived all through time, while never being at more than one place at a given time. This way of looking at things is obviously connected with the complete separation of space and time in which people formerly believed. When we substitute space-time for space and time, we shall naturally expect to derive the physical world from constituents which are as limited in time as in space. Such constituents are what we call 'events.' An event does not persist and move, like the traditional piece of matter; it merely exists for its little moment and then ceases. A piece of matter will thus be resolved into a series of events. Just as, in the old view, an extended body was composed of a number of particles, so, now, each particle, being extended in time, must be regarded as composed of what we may call 'event-particles.' The whole series of these events makes up the whole history of the particle, and the particle is regarded as *being* its history, not some metaphysical entity to which the events happen. This view is rendered necessary by the fact that relativity compels us to place time and space more on a level than they were in the older physics.

This abstract requirement must be brought into relation with the known facts of the physical world. Now what are the known facts? Let us take it as conceded that light consists of waves travelling with the received velocity. We then know a great deal about what goes on in the parts of space-time where there is no matter; we know, that is to say, that there are periodic occurrences (light-waves) obeying certain laws. These light-waves start from atoms, and the modern theory of the structure of the atom enables us to know a great deal about the circumstances under which they start, and the reasons which determine their wave-lengths. We can find out not only how one light-wave travels, but how its source moves relatively to ourselves. But when I say this I am assuming that we can recognize a source of light as the same at two slightly different times. This is, however, the very thing which had to be investigated.

We saw, in the preceding chapter, how a group of connected events can be formed, all related to each other by a law, and all ranged about a centre in space-time. Such a group of events will be the arrival, at various places, of the light-waves emitted by a brief flash of light. We do not need to suppose that anything particular is happening at the centre; certainly we do not need to suppose that we know *what* is happening there. What we know is that, as a matter of geometry, the group of events in question are ranged about a centre, like widening ripples on a pool when a fly has touched it. We can hypothetically invent an occurrence which is to have happened at the centre, and set forth laws by which the consequent disturbance is transmitted. This hypothetical occurrence will then appear to common sense as the 'cause' of the disturbance. It will also count as one event in the biography of the particle of matter which is supposed to occupy the centre of the disturbance.

Now we find not only that one light-wave travels outward from a centre according to a certain law, but also that, in general, it is followed by other closely similar light-waves. The sun, for example, does not change its appearance suddenly; even if a cloud passes across it during a high wind, the transition is gradual, though swift. In this way a group of occurrences connected with a centre at one point of space-time is brought into relation with other very similar groups whose centres are at neighbouring points of space-time. For each of these other groups common sense invents similar hypothetical occurrences to occupy their centres, and says that all these hypothetical occurrences are part of one history; that is to say, it invents a hypothetical 'particle' to which the hypothetical occurrences are to have occurred. It is only by this double use of hypothesis, perfectly unnecessary in each case, that we arrive at anything that can be called 'matter' in the old sense of the word.

If we are to avoid unnecessary hypotheses, we shall say that an atom at a given moment *is* the various disturbances in the surrounding medium which, in ordinary language, would be said to be 'caused' by it. But we shall not take these disturbances at what is, for us, the moment in question, since that would make them depend upon the observer; we shall instead travel outward from the atom with the velocity of light, and take the disturbance we find in each place as we reach it. The closely similar set of disturbances, with very nearly the same centre, which is found existing slightly earlier or slightly later, will be defined as *being* the atom at a slightly earlier or slightly later moment. In this way, we preserve all the laws of physics, without having recourse to unnecessary hypotheses or inferred entities, and we remain in harmony with the general principle of economy which has enabled the theory of relativity to clear away so much useless lumber.

Common sense imagines that when it sees a table it sees a table. This is a gross delusion. When common-sense sees a table, certain light-waves reach its eyes, and these are of a sort which, in its previous experience, has been associated with certain sensations of touch, as well as with other people's testimony that they also saw the table. But none of this ever brought us to the table itself. The light-waves caused occurrences in our eyes, and these caused occurrences in the optic nerve, and these in turn caused occurrences in the brain. Any one of these, happening without the usual preliminaries, would have caused us to have the sensations we call 'seeing the table,' even if there had been no table. (Of course, if matter in general is to be interpreted as a group of occurrences, this must apply also to the eye, the optic nerve and the brain.) As to the sense of touch when we press the table with our fingers, that is an electric disturbance on the electrons and protons of our finger-tips, produced, according to modern physics, by the proximity of the electrons and protons in the table. If the same disturbance in our finger-tips arose in any other way, we should have the sensations, in spite of there being no table. The testimony of others is obviously a second-hand affair. A witness in a law court, if asked whether he had seen some occurrence, would not be allowed to reply that he believed so because of the testimony of others to that effect. In any case, testimony consists of sound-waves and demands psychological as well as physical interpretation; its connection with the object is therefore very indirect. For all these reasons, when we say that a man 'sees a table,' we use a highly abbreviated form of expression, concealing complicated and difficult inferences, the validity of which may well be open to question.

But we are in danger of becoming entangled in psychological questions, which we must avoid if we can. Let us therefore return to the purely physical point of view.

What I wish to suggest may be put as follows. Everything that occurs elsewhere, owing to the existence of an atom, can be explored experimentally, at least in theory, unless it occurs in certain concealed ways. But what occurs within the atom (if anything occurs there) it is absolutely impossible to know: there is no conceivable apparatus by which we could obtain even a glimpse of it. An atom is known by its 'effects.' But the word 'effects' belongs to a view of causation which will not fit modern physics, and in particular will not fit relativity. All that we have a right to say is that certain groups of occurrences happen together, that is to say, in neighbouring parts of space-time. A given observer will regard one member of the group as earlier than the other, but another observer may judge the time-order differently. And even when the time-order is the same for all observers, all that we really have is a connection between the two events, which works equally backwards and forwards. It is not true that the past determines the future in some sense other than that in which the future determines the past: the apparent difference is only due to our ignorance, because we know less about the future than about the past. This is a mere accident: there might be beings who would remember the future and have to infer the past. The feelings of such beings in these matters would be the exact opposite of our own, but no more fallacious.

It seems fairly clear that all the facts and laws of physics can be interpreted without assuming that 'matter' is anything more than groups of events, each event being of the sort which we should naturally regard as 'caused' by the matter in question. This does not involve any change in the symbols or formulae of physics: it is merely a question of interpretation of the symbols.

This latitude in interpretation is a characteristic of mathematical physics. What we know is certain very abstract logical relations, which we express in mathematical formulae; we know also that, at certain points, we arrive at results which are capable of being tested experimentally. Take, for example, the eclipse observations by which Einstein's theory as to the bending of light was established. The actual observation consisted in the careful measurement of certain distances on certain photographic plates. The formulae which were to be verified were concerned with the course of light in passing near the sun. Although the part of these formulae which gives the observed result must always be interpreted in the same way, the other part of them may be capable of a great variety of interpretations. The formulae giving the motions of the planets are almost exactly the same in Einstein's theory as in Newton's, but the meaning of the formulae is quite different. It may be said generally that, in the mathematical treatment of nature, we can be far more certain that our formulae are approximately correct than we can be as to the correctness of this or that interpretation of them. And so in the case with which this chapter is concerned; the question as to the nature of an electron or a proton is by no means answered when we know all that mathematical physics has to say as to the laws of its motion and the laws of its interaction with the environment. A definite and conclusive answer to our question is not possible, just because a variety of answers are compatible with the truth of mathematical physics. Nevertheless some answers are preferable to others, because some have a greater probability in their favour. We have been seeking, in this chapter, to define matter so that there *must* be such a thing, if the formulae of physics are true. If we had made our definition such as to secure that a particle of matter should be what one thinks of as substantial, a hard, definite lump, we should not have been *sure* that any such thing exists. That is why our definition, though it may seem complicated, is preferable from the point of view of logical economy and scientific caution.

CHAPTER XV

Philosophical Consequences

THE philosophical consequences of relativity are neither so great nor so startling as is sometimes thought. It throws very little light on time-honoured controversies, such as that between realism and idealism. Some people think that it supports Kant's view that space and time are 'subjective' and are 'forms of intuition.' I think such people have been misled by the way in which writers on relativity speak of 'the observer.' It is natural to suppose that the observer is a human being, or at least a mind; but he is just as likely to be a photographic plate or a clock. That is to say, the odd results as to the difference between one 'point of view' and another are concerned with 'point of view' in a sense applicable to physical instruments just as much as to people with perceptions. The 'subjectivity' concerned in the theory of relativity is a *physical* subjectivity, which would exist equally if there were no such things as minds or senses in the world.

Moreover, it is a strictly limited subjectivity. The theory does not say that *everything* is relative; on the contrary, it gives a technique for distinguishing what is relative from what belongs to a physical occurrence in its own right. If we are going to say that the theory supports Kant about space and time, we shall have to say that it refutes him about space-time. In my view, neither statement is correct. I see no reason why, on such issues, philosophers should not all stick to the views they previously held. There were no conclusive arguments on either side before, and there are none now; to hold either view shows a dogmatic rather than a scientific temper.

Nevertheless, when the ideas involved in Einstein's work have become familiar, as they will do when they are taught in schools, certain changes in our habits of thought are likely to result, and to have great importance in the long run.

One thing which emerges is that physics tells us much less about the physical world than we thought it did. Almost all the 'great principles' of traditional physics turn out to be like the 'great law' that there are always three feet to a yard; others turn out to be downright false. The conservation of mass may serve to illustrate both these misfortunes to which a 'law' is liable. Mass used to be defined as 'quantity of matter,' and as far as experiment showed it was never increased or diminished. But with the greater accuracy of modern measurements, curious things were found to happen. In the first place, the mass as measured was found to increase with the velocity; this kind of mass was found to be really the same thing as energy. This kind of mass is not constant for a given body. The law itself, however, is to be regarded as a truism, of the nature of the 'law' that there are three feet to a yard; it results from our methods of measurement, and does not express a genuine property of matter. The other kind of mass, which we may call 'proper mass,' is that which is found to be the mass by an observer moving with the body. This is the ordinary terrestrial case where the body we are weighing is not flying through the air. The 'proper mass' of a body is very nearly constant, but not quite. One would suppose that if you have four 1 lb. weights, and you put them all together into the scales, they will together weigh 4 lbs. This is a fond delusion: they weigh rather less, though not enough less to be discovered by even the most careful measurements. In the case of four hydrogen atoms, however, when they are put together to make one helium atom, the defect is noticeable; the helium atom weighs measurably less than four separate hydrogen atoms.

Broadly speaking, traditional physics has collapsed into two portions, truisms and geography.

The world which the theory of relativity presents to our imagination is not so much a world of 'things' in 'motion' as a world of *events*. It is true that there are still particles which seem to persist, but these (as we saw in the preceding chapter) are really to be conceived as strings of connected events, like the successive notes of a song. It is *events* that are the stuff of relativity physics. Between two events which are not too remote from each other there is, in the general theory as in the special theory, a measurable relation called 'interval,' which appears to be the physical reality of which lapse of time and distance in space are two more or less confused representations. Between two distant events, there is not any one definite interval. But there is one way of moving from one event to another which makes the sum of all the little intervals along the route greater than by any other route. This route is called a 'geodesic,' and it is the route which a body will choose if left to itself.

The whole of relativity physics is a much more step-by-step matter than the physics and geometry of former days. Euclid's straight lines have to be replaced by light-rays, which do not quite come up to Euclid's standard of straightness when they pass near the sun or any other very heavy body. The sum of the angles of a triangle is still thought to be two right angles in very small regions of empty space, but not in any extended region. Nowhere can we find a place where Euclid is exactly true. Propositions which used to be proved by reasoning have now become either conventions, or merely approximate truths verified by observation.

It is a curious fact—of which relativity is not the only illustration—that, as reasoning improves, its claims to the power of proving facts grow less and less. Logic used to be thought to teach us how to draw inferences; now, it teaches us rather how not to draw inferences. Animals and children are terribly prone to inference: a horse is surprised beyond measure if you take an unusual turning. When men began to reason, they tried to justify the inferences that they had drawn unthinkingly in earlier days. A great deal of bad philosophy and bad science resulted from this propensity. 'Great principles,' such as the 'uniformity of nature,' the 'law of universal causation,' and so on, are attempts to bolster up our belief that what has often happened before will happen again, which is no better founded than the horse's belief that you will take the turning you usually take. It is not altogether easy to see what is to replace these pseudo-principles in the practice of science; but perhaps the theory of relativity gives us a glimpse of the kind of thing we may expect. Causation, in the old sense, no longer has a place in theoretical physics. There is, of course, something else which takes its place, but the substitute appears to have a better empirical foundation than the old principle which it has superseded.

The collapse of the notion of one all-embracing time, in which all events throughout the universe can be dated, must in the long run affect our views as to cause and effect, evolution, and many other matters. For instance, the question whether, on the whole, there is progress in the universe, may depend upon our choice of a measure of time. If we choose one out of a number of equally good clocks, we may find that the universe is progressing as fast as the most optimistic American thinks it is; if we choose another equally good clock, we may find that the universe is going from bad to worse as fast as the most melancholy Slav could imagine. Thus optimism and pessimism are neither true nor false, but depend upon the choice of clocks.

The effect of this upon a certain type of emotion is devastating. The poet speaks of

> One far-off divine event
> To which the whole creation moves.

But if the event is sufficiently far off, and the creation moves sufficiently quickly, some parts will judge that the event has already happened, while others will judge that it is still in the future. This spoils the poetry. The second line ought to be:

> To which some parts of the creation move, while others move away from it.

But this won't do. I suggest that an emotion which can be destroyed by a little mathematics is neither very genuine nor very valuable. But this line of argument would lead to a criticism of the Victorian age, which lies outside my theme.

What we know about the physical world, I repeat, is much more abstract than was formerly supposed. Between bodies there are occurrences, such as light-waves; of the *laws* of these occurrences we know something—just so much as can be expressed in mathematical formulae—but of their *nature* we know nothing. Of the bodies themselves, as we saw in the preceding chapter, we know so little that we cannot even be sure that they are anything: they *may* be merely groups of events in other places, those events which we should naturally regard as their effects. We naturally interpret the world pictorially; that is to say, we imagine that what goes on is more or less like what we see. But in fact this likeness can only extend to certain formal logical properties expressing structure, so that all we can know is certain general characteristics of its changes. Perhaps an illustration may make the matter clear. Between a piece of orchestral music as played, and the same piece of music as printed in the score, there is a certain resemblance, which may be described as a resemblance of structure. The resemblance is of such a sort that, when you know the rules, you can infer the music from the score or the score from the music. But suppose you had been stone-deaf from birth, but had lived among musical people. You could understand, if you had learned to speak and to do lip-reading, that the musical scores represented something quite different from themselves in intrinsic quality, though similar in structure.[1] The value of music would be completely unimaginable to you, but you could infer all its mathematical characteristics, since they are the same as those of the score. Now our knowledge of nature is something like this. We can read the scores, and infer just so much as our stone-deaf person could have inferred about music. But we have not the advantages which he derived from association with musical people. We cannot know whether the music represented by the scores is beautiful or hideous; perhaps, in the last analysis, we cannot be quite sure that the scores represent anything but themselves. But this is a doubt which the physicist, in his professional capacity, cannot permit himself to entertain.

Assuming the utmost that can be claimed for physics, it does not tell us what it is that changes, or what are its various states; it only tells us such things as that changes follow each other periodically, or spread with a certain speed. Even now we are probably not at the end of the process of stripping away what is merely imagination, in order to reach the core of true

[1] For the definition of 'structure' see the present author's *Introduction to Mathematical Philosophy*.

scientific knowledge. The theory of relativity has accomplished a very great deal in this respect, and in doing so has taken us nearer and nearer to bare structure, which is the mathematician's goal—not because it is the only thing in which he is interested as a human being, but because it is the only thing that he can express in mathematical formulae. But far as we have travelled in the direction of abstraction, it may be that we shall have to travel farther still.

In the preceding chapter, I suggested what may be called a minimum definition of matter, that is to say, one in which matter has, so to speak, as little 'substance' as is compatible with the truth of physics. In adopting a definition of this kind, we are playing for safety: our tenuous matter will exist, even if something more beefy also exists. We tried to make our definition of matter, like Isabella's gruel in Jane Austen, 'thin, but not too thin.' We shall, however, fall into error if we assert positively that matter is nothing more than this. Leibniz thought that a piece of matter is really a colony of souls. There is nothing to show that he was wrong, though there is also nothing to show that he was right: we know no more about it either way than we do about the flora and fauna of Mars.

To the non-mathematical mind, the abstract character of our physical knowledge may seem unsatisfactory. From an artistic or imaginative point of view, it is perhaps regrettable, but from a practical point of view it is of no consequence. Abstraction, difficult as it is, is the source of practical power. A financier, whose dealings with the world are more abstract than those of any other 'practical' man, is also more powerful than any other practical man. He can deal in wheat or cotton without needing ever to have seen either: all he needs to know is whether they will go up or down. This is abstract mathematical knowledge, at least as compared to the knowledge of the agriculturist. Similarly the physicist, who knows nothing of matter except certain laws of its movements, nevertheless knows enough to enable him to manipulate it. After working through whole strings of equations, in which the symbols stand for things whose intrinsic nature can never be known to us, he arrives at last at a result which can be interpreted in terms of our own perceptions, and utilized to bring about desired effects in our own lives. What we know about matter, abstract and schematic as it is, is enough, in principle, to tell us the rules according to which it produces perceptions and feelings in ourselves; and it is upon these rules that the *practical* uses of physics depend.

The final conclusion is that we know very little, and yet it is astonishing that we know so much, and still more astonishing that so little knowledge can give us so much power.

GEORGE ALLEN & UNWIN LTD

Head Office:
London: 40 Museum Street, W.C.1

Trade orders and enquiries
Park Lane, Hemel Hempstead, Herts.

Auckland: P.O. Box 36012, Northcote Central N.4
Barbados: P.O. Box 222, Bridgetown
Bombay: 15 Graham Road, Ballard Estate, Bombay 1
Buenos Aires: Escritorio 454-459, Florida 165
Beirut: Deeb Building, Jeanne d'Arc Street
Calcutta: 17 Chittaranjan Avenue, Calcutta 13
Cape Town, 68 Shortmarket Street
Hong Kong: 105 Wing On Mansion, 26 Hancow Road, Kowloon
Ibadan: P.O. Box 62
Karachi: Karachi Chambers, McLeod Road
Madras: Mohan Mansions, 38c Mount Road, Madras 6
Mexico: Villalongin 32, Mexico 5, D.F.
Nairobi: P.O. Box 30583
New Delhi: 13-14 Asaf Ali Road, New Delhi 1
Philippines: P.O. Box 157, Quezon City D-502
Rio de Janeiro: Caixa Postal 2537-Zc-00
Singapore: 34c Prinsep Street, Singapore 7
Sydney N.S.W.: Bradbury House, 55 York Street
Tokyo: P.O. Box 26, Kamata
Toronto: 81 Curlew Drive, Don Mills

BERTRAND RUSSELL

HUMAN KNOWLEDGE: ITS SCOPE AND LIMITS

This book is intended for the general reader, not for professional philosophers. It begins with a brief survey of what science professes to know about the universe. In this survey the attempt is to be as far as possible impartial and impersonal; the aim is to come as near as our capacities permit to describing the world as it might appear to an observer of miraculous perceptive powers viewing it from without. In science, we are concerned with what we *know* rather than what *we* know. We attempt to use an order in our description which ignores, for the moment, the fact that we are part of the universe, and that any account which we can give of it depends upon its effects upon ourselves, and is to this extent inevitably anthropocentric.

Bertrand Russell accordingly begins with the system of galaxies, and passes on, by stages, to our own galaxy, our own little solar system, our own tiny planet, the infinitesimal specks of life upon its surface, and finally, as the climax of insignificance, the bodies and minds of those odd beings that imagine themselves the lords of creation and the end of the whole vast cosmos.

But this survey, which seems to end in the pettiness of Man and all his concerns, is only one side of the truth. There is another side, which must be brought out by a survey of a different kind. In this second kind of survey the question is no longer what the universe is, but how we come to know whatever we do know about it. In this survey Man again occupies the centre, as in the Ptolemaic astronomy. What we know of the world we know by means of events in our own lives, events which, but for the power of thought, would remain merely private.

The book inquires what are our data, and what are the principles by means of which we make our inferences. The data from which these inferences proceed are private to ourselves; what we call 'seeing the sun' is an event in the life of the seer, from which the astronomer's sun has to be inferred by a long and elaborate process. It is evident that, if the world were a higgledy-piggledy chaos, inferences of this kind would be impossible; but for causal inter-connectedness, what happens in one place would afford no indication of what has happened in another. It is the process from private sensation and thought to impersonal science that forms the chief topic of the book. The road is at times difficult, but until we have traversed it neither the scope nor the limitations of human knowledge can be adequately understood.

Muirhead Library of Philosophy

BERTRAND RUSSELL

OUR KNOWLEDGE OF THE EXTERNAL WORLD

The lectures in this volume are designed to show, by means of examples, the nature, capacity, and limitations of the logical analytic method in philosophy.

'This brilliant, lucid and amusing book which . . . everyone can understand.'—*New Statesman*

'It is in every sense an epoch making book; one that has been needed and expected for years.'—*Cambridge Magazine*

'The author maintains the fresh and brilliant yet easy style which always makes his writings a pleasure to read.'—*Nature*

AN OUTLINE OF PHILOSOPHY

Contents include: Philosophic Doubts; Man from Without: Man and His Environment; The Process of Learning in Animals and Infants; Language; Perception Objectively Regarded; Memory Objectively Regarded; Knowledge Behaviouristically Considered; The Physical World; The Structure of the Atom; Physics and Perception; Physical and Perpetual Space; Man from Within: Self Observation; Images; Imagination and Memory; Consciousness? Emotion, Desire, and Will; Some Great Philosophers of the Past; Truth and Falsehood; Events, Matter and Mind; Man's Place in the Universe.

ANALYSIS OF MIND

From the tendency of many psychologists to adopt an essentially materialistic position, while the physicists, especially Einstein, have been making 'matter' less and less material, the author has developed a view reconciling the two conflicting tendencies, according to which the 'stuff' of the world is neither mental nor material, but a 'neutral stuff' out of which both are constructed. 'A most brilliant essay in psychology.'—*New Statesman and Nation*

Muirhead Library of Philosophy

BERTRAND RUSSELL: THE PASSIONATE SCEPTIC

by Alan Wood

This book is the first of its kind to be published about one of the only two present-day Englishmen certain to be remembered a thousand years hence. It is inspired by the author's belief that a portrait drawn from life, whatever its shortcomings, can have an authenticity of detail unattainable by the orthodox scholarship which waits till a great man is dead before trying to get to know him and discuss him.

Several years ago Alan Wood began research for his forthcoming book, *Russell's Philosophy: A Study of its Development*. The present book, mainly for the general reader, contains some chapters discussing Russell's philosophy. But it also considers his views on politics, pacifism, marriage, and education; mentions his travels in Germany, Russia, China, and Australia; tells something of his imprisonment in 1918, his experiences in running an advanced school in England, and his dismissal from a New York Professorship in 1940. The outline given of Russell's turbulent career, combining supreme achievement with continual battles against misfortune, provides many unsuspected sidelights on Russell as a man, and on his rare blend of austere logical precision, fierce human compassion, and irresistible wit. Mr. Wood has not only secured new first-hand information about Russell's way of working as philosopher, but also about such points as the reason why he told the Governor of Brixton Prison that he wanted an ourang-outang, and his reason for praising a gasometer at Oxford.

'. . . provocative and genial biography. . . . This brilliantly entertaining book compliments its central figure better by an entire freedom from humbug.'—*Daily Telegraph*

'Mr. Wood's biography catches the authentic spirit of a life packed with mental adventure. Minor blemishes do not seriously detract from its absorbing interest.'—*The Times*

'. . . a brilliant and oddly moving biography. . . . Mr. Wood's is a remarkably candid yet always affectionate and good-mannered portrait of one of the most remarkable men of our day.'—*Evening News*

GEORGE ALLEN & UNWIN LTD